Prosperity

Charles Fillmore

IAP © 2009

Copyright © 2009 by IAP. Las Vegas, Nevada, USA.

All rights reserved. This publication is protected by Copyright and written permission should be obtained from the publisher prior to any prohibited reproduction, storage or transmission.

Printed in Scotts Valley, CA – USA.

Fillmore, Charles.

Prosperity / Charles Fillmore – 1st ed.

   1. Self-help   2. Religion

Book Cover: ©IAP

Contents

Foreword	4
Lesson 1 Spiritual Substance, the Fundamental Basis of the Universe	6
Lesson 2 Spiritual Mind, the Omnipresent Directive Principle of Prosperity	15
Lesson 3 Faith in the Invisible Substance, the Key to Demonstration	24
Lesson 4 Man, the Inlet and Outlet of Divine Mind	32
Lesson 5 The Law That Governs the Manifestation of Supply	40
Lesson 6 Wealth of Mind Expresses Itself in Riches	49
Lesson 7 God Has Provided Prosperity for Every Home	57
Lesson 8 God Will Pay Your Debts	65
Lesson 9 Tithing, the Road to Prosperity	73
Lesson 10 Right Giving, the Key to Abundant Receiving	81
Lesson 11 Laying Up Treasures	89
Lesson 12 Overcoming the Thought of Lack	97
Question Helps	105

Foreword

It is perfectly logical to assume that a wise and competent Creator would provide for the needs of His creatures in their various stages of growth. The supply would be given as required and as the necessary effort for its appropriation was made by the creature. Temporal needs would be met by temporal things, mental needs by things of like character, and spiritual needs by spiritual elements. For simplification of distribution all would be composed of one primal spiritual substance, which under proper direction could be transformed into all products at the will of the operator. This is a crude yet true illustration of the underlying principles on which the human family is supplied on this earth. The Father has provided a universal seed substance that responds with magical power to the active mind of man. Faith in the increasing capacity of this seed substance, whether wrapped in visible husks or latent in invisible electrical units, always rewards man with the fruits of his labor.

The farmer may seem to get his supply from the seeds he plants, but he would never plant a seed unless he had faith in its innate capacity to increase, and that seed would never multiply without the quickening life of Spirit. Thus we see that all increase of substance depends on the quickening life of Spirit, and this fact gives us the key to mental processes that when used spiritually will greatly increase and at the same time simplify our appropriation of that inexhaustible substance which creative Mind has so generously provided.

In the following lessons we have attempted to explain man's lawful appropriation of the supplies spiritually and electrically provided by God. When we understand and adjust our mind to the realm or kingdom where these rich ideas and their electrical thought forms exist we shall experience in our temporal affairs what is called "prosperity."

We said "their electrical thought forms." Let us explain that all creative processes involve a realm of ideas and a realm of patterns or expressions of those ideas. The patterns arrest or "bottle up" the free electric units that sustain the visible thing. Thus creation is in its processes a trinity, and back of the visible universe are both the original creative idea and the cosmic rays that crystallize into earthly things. When we understand this trinity in its various activities we shall be able to reconcile the discoveries of modern science with the fundamentals of religion.

Modern science teaches us that space is heavily charged with energies that would transform the earth if they could be controlled. Sir Oliver Lodge says that a single cubic inch of the ether contains energy enough to run a forty-horse-power engine forty million years. The divergence of opinion among physicists as to the reality of the ether does not nullify the existence in space of tremendous potentialities. Sir Arthur Eddington says that about half the leading physicists assert that the ether exists and the other half deny its existence, but, in his words, "both parties mean exactly the same thing, and are divided only by words."

Spiritual understanding says that the ether exists as an emanation of mind and should not be confused in its limitations with matter. Mathematical measurements applied to the ether work it out of existence because its reality is in the Mind that conceived it and its being is governed and sustained by ideas, and ideas have no physical dimensions. So the ether will have existence and deposit matter only so long as Mind has use for it. When infinite Mind has completed the cycles of creation, both the invisible and the visible universes will be rolled up as a scroll and disappear and only Mind remain. "And all the host of heaven shall be dissolved, and the heavens shall be rolled together as a scroll; and all their host shall fade away."

It adds greatly to the stability of a Christian's faith to know that Jesus anticipated the discoveries of modern science of the existence of that kingdom called "the ether." He named it the kingdom of the heavens, and His illustrations of its possibilities are unsurpassed. He did not say it was a place the good would inherit after death but an estate we could have here and now. "It is your Father's good pleasure to give you the kingdom."

Jesus taught that we can incorporate life-giving rays into our mind, body, and affairs through faith. Where physicists merely describe the mechanical presence of life as energy, Jesus taught man how by the exercise of his mind he can make that life obey

him. Instead of a universe of blind mechanical forces Jesus showed the universe to be persuaded and directed by intelligence.

What we need to realize above all else is that God has provided for the most minute needs of our daily life and that if we lack anything it is because we have not used our mind in making the right contact with the supermind and the cosmic ray that automatically flows from it

Lesson 1
Spiritual Substance, the Fundamental Basis of the Universe

Divine mind is the one and only reality. When we incorporate the ideas that form this Mind into our mind and persevere in those ideas, a mighty strength wells up within us. Then we have a foundation for the spiritual body, the body not made with hands, eternal in the heavens. When the spiritual body is established in consciousness, its strength and power is transmitted to the visible body and to all the things that we touch in the world about us.

Spiritual discernment reveals that we are now in the dawn of a new era, that the old methods of supply and support are fast passing away, and that new methods are waiting to be brought forth. In the coming commerce man will not be a slave to money. Humanity's daily needs will be met in ways that are not now thought practical. We shall serve for the joy of serving, and prosperity will flow to us and through us in streams of plenty. The supply and support that love and zeal will set in motion are not as yet largely used by man, but those who have tested their providing power are loud in their praise.

The dynamic power of the supermind in man has been sporadically displayed by men and women of every nation. It is usually connected with some religious rite in which mystery and priestly authority prevail. The so-called "common herd" are kept in darkness with respect to the source of the superhuman power of occult adepts and holy men. But we have seen a "great light" in the discovery by physical scientists that the atom conceals electronic energies whose mathematical arrangement determines the character of all the fundamental elements of nature. This discovery has disrupted the science based on the old mechanical atomic theory, but has also given Christian metaphysicians a new understanding of the dynamics back of Spirit.

Science now postulates space rather than matter as the source of life. It says that the very air is alive with dynamic forces that await man's grasp and utilization and that these invisible, omnipresent energies possess potentialities far beyond our most exalted conception. What we have been taught about the glories of heaven pales into insignificance compared with the glories of the radiant rays--popularly referred to as the "ether." We are told by science that we have utilized very meagerly this mighty ocean of ether in producing from it the light and power of electricity. The seemingly tremendous force generated by the whirl of our dynamos is but a weak dribble from a universe of energy. The invisible waves that carry radio programs everywhere are but a mere hint of an intelligent power that penetrates and permeates every germ of life, visible and invisible. Scientific minds the world over have been tremendously moved by these revolutionary discoveries, and they have not found language adequate to explain their magnitude. Although a number of books have been written by scientists, setting forth guardedly the far-reaching effects that will inevitably follow man's appropriation of the easily accessible ether, none has dared to tell the whole story. The fact is that the greatest discovery of all ages is that of physical science that all things apparently have their source in the invisible, intangible ether. What Jesus taught so profoundly in symbols about the riches of the kingdom of the heavens has now been proved true.

According to the Greek, the language in which the New Testament has come down to us, Jesus did not use the word heaven but the word heavens in His teaching. He was not telling us of the glories of some faraway place called "heaven" but was revealing the properties of the "heavens" all around us, called both "space" and "ether" by physicists. He taught not only its dynamic but also its intelligent character, and said that the entity that rules it is within man: "The kingdom of God is within you." He not only described this kingdom of the heavens in numerous parables but made its attainment by man the greatest object of human existence. He not only set this as man's goal but attained it Himself, thereby demonstrating that His teaching is practical as well as true.

The scientists tell us that the ether is charged with electricity, magnetism, light rays, X rays, cosmic rays, and other dynamic radiations; that it is the source of all life, light, heat, energy, gravitation, attraction, repulsion; in short, that it is the interpenetrating essence of everything that exists on the earth. In other words, science gives to the ether all the attractions of heaven without directly saying so. Jesus epitomized the subject when He told His followers that it was the kingdom from which God clothed and fed all His children. "Seek ye first his kingdom, and his righteousness; and all these things shall be added unto you." Science says that the electrical particles that break into light in our earth's atmosphere are also a source of all substance and matter. Jesus said that He was the substance and bread that came from the heavens. When will our civilization begin really to appropriate and use this mighty ocean of substance and life spiritually as well as physically?

This inexhaustible mind substance is available at all times and in all places to those who have learned to lay hold of it in consciousness. The simplest, shortest, and most direct way of doing this was explained when Jesus said, "Whosoever ... shall not doubt in his heart, but shall believe that what he saith cometh to pass, he shall have it." When we know that certain potent ideas exist in the invisible mind expressions, named by science both "ether" and "space" and that we have been provided with the mind to lay hold of them, it is easy to put the law into action through thought and word and deed.

"There is a tide in the affairs of men,

Which, taken at the flood, leads on to fortune," said Shakespeare. That flood tide awaits us in the cosmic spaces, the paradise of God.

The spiritual substance from which comes all visible wealth is never depleted. It is right with you all the time and responds to your faith in it and your demands on it. It is not affected by our ignorant talk of hard times, though we are affected because our thoughts and words govern our demonstration. The unfailing resource is always ready to give. It has no choice in the matter; it must give, for that is its nature. Pour your living words of faith into the omnipresent substance, and you will be prospered though all the banks in the world close their doors. Turn the great energy of your thinking toward "plenty" ideas, and you will have plenty regardless of what men about you are saying or doing.

God is substance, but if by this statement we mean that God is matter, a thing of time or condition, then we should say that God is substanceless. God is not confined to that form of substance which we term matter. God is the intangible essence of that which man has formed into and named matter. Matter is a mental limitation of that divine substance whose vital and inherent character is manifest in all life expression.

God substance may be conceived as God energy, or Spirit light, and "God said, let there be light, and there was light." This is in harmony with the conclusions of some of the most advanced physicists. Sir James Jeans says, in "The Mysterious Universe," "The tendency of modern physics is to resolve the whole material universe into waves, and nothing but waves. These waves are of two kinds: bottled-up waves, which we call matter, and unbottled waves, which we call radiation, or light. The process of annihilation of matter is merely unbottling imprisoned wave energy, and setting it free to travel through space."

Spirit is not matter. Spirit is not person. In order to perceive the essence of Being we must drop from our mind all thought that God is in any way circumscribed or has any of the limitations that we associate with things or persons having form or shape. "Thou shalt not make unto thee a graven image, nor any likeness of any thing that is in heaven above, or that is in the earth beneath."

God is substance, not matter, because matter is formed, while God is the formless. God substance lies back of matter and form. It is the basis of all form yet does not enter into any form as a finality. Substance cannot be seen, touched, tasted, or smelled, yet it is more substantial than matter, for it is the only substantiality in the universe. Its nature is to "sub-stand" or "stand under" or behind matter as its support and only reality.

Job says, "The Almighty shall be thy defense, and thou shalt have plenty of silver." This refers to universal substance, for silver and gold are manifestations of an everywhere present substance and are used as symbols for it. Lew Wallace, in "Ben-Hur," refers to the kingdom as "beaten gold." You have doubtless in your own experience caught sight of this everywhere present substance in your silence, when it seemed like golden snowflakes falling all about you. This was the first manifestation from the overflow of the universal substance in your consciousness.

Substance is first given form in the mind, and as it becomes manifest it goes through a threefold activity. In laying hold of substance in the mind and bringing it into manifestation, we play a most important part. We do it according to our decree. "Thou shalt decree a thing, and it shall be established unto thee." We are always decreeing, sometimes consciously, often unconsciously, and with every thought and word we are increasing or diminishing the threefold activity of substance. The resulting manifestation conforms to our thought, "As he thinketh within himself, so is he."

There is no scarcity of the air you breathe. There is plenty of air, all you will ever need, but if you close your lungs and refuse to breathe, you will not get it and may suffocate for lack of air. When you recognize the presence of abundance of air and open your lungs to breathe it deeply, you get a larger inspiration. This is exactly what you should do with your mind in regard to substance. There is an all-sufficiency of all things, just as there is an all-sufficiency of air. The only lack is our own lack of appropriation. We must seek the kingdom of God and appropriate it aright before things will be added to us in fullness.

There is a kingdom of abundance of all things, and it may be found by those who seek it and are willing to comply with its laws. Jesus said that it is hard for a rich man to enter into the kingdom of heaven. This does not mean that it is hard because of his wealth, for the poor man gets in no faster and no easier. It is not money but the thoughts men hold about money, its source, its ownership, and its use, that keep them out of the kingdom. Men's thoughts about money are like their thoughts about all possessions; they believe that things coming out of the earth are theirs to claim and control as individual property, and may be hoarded away and depended on, regardless of how much other men may be in need of them. The same belief is prevalent among both rich and poor, and even if the two classes were suddenly to change places, the inequalities of wealth would not be remedied. Only a fundamental change in the thoughts of wealth could do that.

Before there is any fundamental social or economic change men must begin to understand their relationship to God and to one another as common heirs to the universal resource that is sufficient for all. They must give up some of their erroneous ideas about their "rights." They must learn that they cannot possess and lock up that which belongs to God without themselves suffering the effects of that sequestration. The poor man is not the greatest sufferer in this concentration of wealth, for he has not concentrated his faith in material things and chained his soul to them. Those who are rich in the things of this world are by their dependence on those things binding themselves to material things and are in material darkness.

Every thought of personal possession must be dropped out of mind before men can come into the realization of the invisible supply. They cannot possess money, houses, or land selfishly, because they cannot possess the universal ideas for which these symbols stand. No man can possess any idea as his own permanently. He may possess its material symbol for a little time on the plane of phenomena, but it is such riches that "moth and rust consume, and where thieves break through and steal."

Men possess as valuables their education, trade, ability, or intellectual talent. Ministers of the gospel possess scholarship or eloquence, and take pride in these spiritual possessions. Yet even these are burdens that must be unloaded before they may enter the kingdom of the heavens. The saint who is puffed up with his saintly goodness must unload his vanity before he gets in. Whoever is ambitious to do good, to excel his fellow men in righteousness, must lose his ambition and desire before he beholds the face of the all-providing Father.

The realm of causes may be compared to steam in a glass boiler. If the glass is clear one may look right at it and see nothing at all. Yet when an escape valve is touched the steam rushes out, condenses and becomes visible. But in this process it has also lost its power. Substance exists in a realm of ideas and is powerful when handled by one who is familiar with its characteristics. The ignorant open the valves of the mind and let ideas flow out into a realm with which they have nothing in common. The powerful ideas of substance are condensed into thoughts of time and space, which ignorance conceives as being necessary to their fruition. Thus their power is lost, and a weary round of seedtime and harvest is inaugurated to fulfill the demands of the world.

It is the mind that believes in personal possessions that limits the full idea. God's world is a world of results that sequentially follow demands. It is in this kingdom that man finds his true home. Labor has ceased for him who has found this inner kingdom. Divine supply is brought forth without laborious struggle: to desire is to have fulfillment.

This is the second step in demonstration for the one who has fully dedicated himself to the divine guidance. He immediately enters into easier experiences and more happiness than the world affords, when he covenants to follow only the good. There is an advanced degree along the same line of initiation into the mysteries of the divine. Before this step may be taken, a deeper and more thorough mental cleansing must be undergone. A higher set of faculties is then awakened within the body, and new avenues of expression are opened for the powers of the Spirit, not only in the body but also in the affairs of the individual. As he proceeds to exercise these faculties he may find some of them clogged by the crystals of dead thought that some selfish ideas have deposited, which makes him go through a fresh cleansing. If he is obedient to the Spirit and willing to follow without cavil or protest, the way is easy for him. If however he questions and argues, as did Job, he will meet many obstructions and his journey will be long and tedious.

Again, he who seeks the kingdom of substance for the sake of the loaves and fishes he may get out of it will surely be disappointed in the end. He may get the loaves and fishes, that is quite possible; but if there remains in his soul any desire to use them for selfish ends, the ultimate result will be disastrous.

Many people are seeking the aid of Spirit to heal them of their physical ills. They have no desire for the higher life, but having found their lusts and passions curtailed by physical infirmities, they want these erased in order that they may continue in their fleshly way. It is the experience of all who have dealt with Spirit that it is a vigorous bodily stimulant. It restores the vitality of the body until it is even more sensitive to pleasure or pain than it was before the spiritual quickening. This supersensitiveness makes it more susceptible and liable to more rapid waste if further indulgence is gratified. That is why those who receive spiritual treatment should be fully instructed in the Truth of Being. They should be shown that the indulgence of bodily passions is a sin against their success in every walk of life and especially in the way of finances and prosperity. If substance is dissipated, every kind of lack begins to be felt. Retribution always follows the indulgence of appetite and passion for mere sensation. Both sinners and saints suffer in this valley of folly. The alternative is to dedicate yourself to the Father's business. Make a definite and detailed covenant with the Father, lay your desires, appetites, and passions at His feet and agree to use all your substance in the most exalted way. Then you are seeking the kingdom, and all things else shall be added unto you.

We want to make this substance that faith has brought to our mind enduring and abiding, so that we do not lose it when banks fail or men talk of "hard times." We must have in our finances a consciousness of the permanency of the omnipresent substance as it abides in us. Some wealthy families succeed in holding their wealth while others dissipate it in one generation because they do not have the consciousness of abiding substance. For many of us there is either a feast or a famine in the matter of money and we need the abiding consciousness. There is no reason why we should not have a continuous even flow of substance both in income and outgo. If we have freely received we must also freely give and keep substance going, confident in our understanding that our supply is unlimited and that it is always right at hand in the omnipresent Mind of God.

In this understanding we can stand "the slings and arrows of outrageous fortune," depressions, losses, and financial failures and still see God as abundant substance waiting to come into manifestation. That is what Paul meant by taking up "the whole armor of God that ye may be able to withstand in the evil day." The substance that has in the past been manifest in our affairs is still here. It is the same substance and it cannot be taken away. Even though there seems to be material lack, there is plenty of substance for all. We are standing in the very midst of it. Like the fish we might ask, "Where is the water," when we live and move and have our being in it. It is in the water, in the air everywhere, abounding, glorious spiritual substance. Take that thought and hold it. Refuse to be shaken from your spiritual stand in the very midst of God's prosperity and plenty, and supply will begin to come forth from the ether and plenty will become more and more manifest in your affairs.

Jesus was so charged with spiritual substance that when the woman touched His garment the healing virtue went out from it and she was healed. There were thousands of people in the crowd, but only the woman who had faith in that substance got it. It was already established in her consciousness, and she knew that her needs would be met if she could make the contact. In this there is a lesson for us. We know that strength is manifest everywhere, for we see it in the mechanical world. A great locomotive starts from the depot, moving slowly at first, but when it gains momentum it speeds down the track like a streak. Thus it is with spiritual strength. Beginning sometimes with a very small

thought, it takes on momentum and eventually becomes a powerful idea. Every one of us can strengthen his hold on the thought of divine substance until it becomes a powerful idea, filling the consciousness and manifesting itself as plenty in all our affairs.

As you lay hold of substance with your mind, make it permanent and enduring. Realize your oneness with it. You are unified with the one living substance, which is God, your all-sufficiency. From this substance you were created; in it you live and move and have your being; by it you are fed and prospered.

The spiritual substance is steadfast and immovable, enduring. It does not fluctuate with market reports. It does not decrease in "hard times" nor increase in "good times." It cannot be hoarded away to cause a deficiency in supply and a higher price. It cannot be exhausted in doles to meet the needs of privation. It is ever the same, constant, abundant, freely circulating and available.

The spiritual substance is a living thing, not an inanimate accumulation of bread that does not satisfy hunger nor water that fails to quench thirst. It is living bread and living water, and he that feeds on God's substance shall never hunger and never thirst. The substance is an abiding thing, not a bank deposit that can be withdrawn nor a fortune that can be lost. It is an unfailing principle that is as sure in its workings as the laws of mathematics. Man can no more be separated from his supply of substance than life can be separated from its source. As God permeates the universe and life permeates every cell of the body, so does substance flow freely through man, free from all limit or qualification.

In the new era that is even now at its dawn we shall have a spirit of prosperity. This principle of the universal substance will be known and acted on, and there will be no place for lack. Supply will be more equalized. There will not be millions of bushels of wheat stored in musty warehouses while people go hungry. There will be no overproduction or underconsumption or other inequalities of supply, for God's substance will be recognized and used by all people. Men will not pile up fortunes one day and lose them the next, for they will no longer fear the integrity of their neighbors nor try to keep their neighbor's share from him.

Is this an impractical utopia? The answer depends on you. Just as soon as you individually recognize the omnipresent substance and put your faith in it, you can look for others around you to do the same. "A little leaven leaveneth the whole lump," and even one life that bears witness to the truth of the prosperity law will quicken the consciousness of the whole community.

Whoever you are and whatever your immediate need, you can demonstrate the law. If your thoughts are confused, become still and know. Be still and know that you are one with the substance and with the law of its manifestation. Say with conviction:

I am strong, immovable Spirit substance.

This will open the door of your mind to an inflow of substance-filled ideas. As they come, use them freely. Do not hesitate or doubt that they will bring results. They are God's ideas given to you in answer to your prayer and in order to supply your needs. They are substance, intelligent, loving, eager to manifest themselves to meet your need.

God is the source of a mighty stream of substance, and you are a tributary of that stream, a channel of expression. Blessing the substance increases its flow. If your money supply is low or your purse seems empty, take it in your hands and bless it. See it filled with the living substance ready to become manifest. As you prepare your meals bless the food with the thought of spiritual substance. When you dress, bless your garments and realize that you are being constantly clothed with God's substance. Do not center your thought on yourself, your interests, your gains or losses, but realize the universal nature of substance. The more conscious you become of the presence of the living substance the more it will manifest itself for you and the richer will be the common good of all.

Do not take anyone's word for it, but try the law for yourself. The other fellow's realization of substance will not guarantee your supply. You must become conscious of it for yourself. Identify yourself with substance until you make it yours; it will change your finances, destroy your fears, stop your worries, and you will soon begin to rejoice in the ever-present bounty of God.

Be still and turn within to the great source. See with the eye of faith that the whole world is filled with substance. See it falling all about you as snowflakes of gold and silver and affirm with assurance:

Jesus Christ is now here raising me to His consciousness of the omnipresent, all-providing God substance, and my prosperity is assured.

I have unbounded faith in the all-present spiritual substance increasing and multiplying at my word.

Lesson 2
Spiritual Mind, the Omnipresent Directive Principle of Prosperity

Everything that appears in the universe had its origin in mind. Mind evolves ideas, and ideas express themselves through thoughts and words. Understanding that ideas have a permanent existence and that they evolve thoughts and words, we see how futile is any attempted reform that does not take them into consideration. This is why legislation and external rules of action are so weak and transient as reforms.

Ideas generate thought currents, as a fire under a boiler generates steam. The idea is the most important factor in every act and must be given first place in our attention if we would bring about any results of a permanent character. Men formulate thoughts and thoughts move the world.

Ideas are centers of consciousness. They have a positive and a negative pole and generate thoughts of every conceivable kind. Hence a man's body, health, intelligence, finances, in fact everything about him, are derived from the ideas to which he gives his attention.

Man has never had a desire that could not somewhere, in the providence of God, be fulfilled. If this were not true, the universe would be weak at its most vital point. Desire is the onward impulse of the ever-evolving soul. It builds from within outward and carries its fulfillment with it as a necessary corollary.

All is mind. Then the things that appear must be expressions of mind. Thus mind is reality, and it also appears as phenomena. The is-ness of mind is but one side of it. Being is not limited to the level of is-ness; it has all possibilities, including that of breaking forth from its inherencies into the realm of appearances. Mind has these two sides, being and appearance, the visible and the invisible. To say that mind is all and yet deny that things do appear to have any place in the allness is to state but half the truth.

An idea is capable of statement as a proposition. The statement is made in response to a desire to know experimentally whether the proposition is capable of proof. A number of elements are involved in the statement of a proposition that are not integral parts of the proposition itself but necessary to its working out. In the simplest mathematical problem processes are used that are not preserved after the problem is solved yet that are necessary to its solution. The figures by the use of which we arrived at the solution are immediately forgotten, but they could not be dispensed with and it is to them we owe the outcome. The exact outcome of each step in the solution is a matter of experiment. The intermediate steps may be changed or retracted many times, but ultimately the problem is solved and the fulfillment of the desired result attained. If this is true of the simplest problem in arithmetic it is equally true of the creation of the universe. "As above, so below." Here is where many who have caught sight of the perfection and wholeness of the ideal fail to demonstrate. They deny the appearance because it does not express perfection in its wholeness.

The student in the depths of a mathematical problem who should judge thus would erase all his figures because the answer was not at once apparent, though he may have already completed a good part of the process leading up to the desired answer. We would not say that a farmer is wise who cuts down his corn in the tassel because it does not show the ripened ears. Do not jump to conclusions. Study a situation carefully in its various aspects before you decide. Consider both sides, the visible and the invisible, the within and the without.

The very fact that you have an ideal condition or world in your mind carries with it the possibility of its fulfillment in expression. Being cannot shirk expression. To think is to express yourself, and you are constantly thinking. You may deny that the things of the outer world have existence, yet as long as you live in contact with them you are recognizing them. When you affirm being and deny the expression of being, you are a "house divided against itself."

We have all wondered why we do not understand more truth than we do or why it is necessary to understand at all, since God is all-wise and all-present. Understanding is one of the essential parts of your I AM identity. Man is a focal point in God consciousness and expresses God. Therefore he must understand the processes that bring about that expression. Infinite Mind is here with all its ideas as a resource for man, and what we are or become is the result of our efforts to accumulate in our own consciousness all the attributes of infinite Mind. We have learned that we can accumulate ideas of power, strength, life, love, and plenty. How should we use these ideas or bring them into outer expression without understanding? Where shall we get this understanding save from the source of all ideas, the one Mind? "But if any of you lacketh wisdom, let him ask of God, who giveth to all liberally and upbraideth not; and it shall be given him."

In following the principles of mathematics we use rules. There is a rule of addition that we must observe when we add; other rules that must be followed when we subtract or multiply. The ideas of Divine Mind can only be expressed when we follow the rules or laws of mind, and these rules require understanding if we would follow them intelligently and achieve results. Man is given all power and authority over all the ideas of infinite Mind, and the idea of wisdom is one of them.

Closely associated with the idea of wisdom in Divine Mind is the idea of love. These ideas are the positive and the negative pole of the creative Principle. "Male and female created he them." The ideas of God-Mind are expressed through the conjunction of wisdom and love. God commanded that these two ideas should be fruitful and multiply and replenish the whole earth with thoughts in expression.

We have access to the divine realm from which all thoughts are projected into the world. We are constantly taking ideas from the spiritual world and forming them into our own conception of the things we desire. Sometimes the finished product does not satisfy or please us. That is because we have taken the idea away from its true parents, wisdom and love, and let it grow to maturity in an atmosphere of error and ignorance.

In the matter of money or riches we have taken the idea of pure substance from the spiritual realm, then have forgotten the substance idea and tried to work it out in a material atmosphere of thought. It was a wonderful idea, but when we took it away from its spiritual parents wisdom and love, it became an unruly and disappointing child. Even if without love and understanding of substance you accumulate gold and silver, your store will not be stable or permanent. It will fluctuate and cause you worry and grief. There are many people who "don't know the value of a dollar," with whom money comes and goes, who are rich today and poor tomorrow. They have no understanding of the substance that is the underlying reality of all wealth.

To have adequate supply at all times, an even flow that is never enough to become a burden yet always enough to meet every demand, we must make union with the Spirit that knows how to handle ideas as substance. Men have the idea that material substance is limited, and they engage in competition trying to grab one another's money. Divine Mind has ideas of substance as unlimited and everywhere present, equally available to all. Since man's work is to express substance ideas in material form, we must find a way to connect ideas of substance with ideas of material expression, to adjust the ideas of man's mind with the ideas of Divine Mind. This is accomplished by faith through prayer.

That part of the Lord's Prayer which reads, "Give us this day our daily bread," is more correctly translated, "Give us today the substance of tomorrow's bread." By prayer we accumulate in our mind ideas of God as the substance of our supply and support. There is no lack of this substance in infinite Mind. Regardless of how much God gives, there is always an abundance left. God does not give us material things, but Mind substance--not money but ideas--ideas that set spiritual forces in motion so that things begin to come to us by the application of the law.

It may be that you solve your financial problem in your dreams. Men often think over their problems just before going to sleep and get a solution in their dreams or immediately upon awakening. This is because their minds were so active on the intellectual plane that they could not make contact with the silent inner plane where ideas work. When the conscious mind is stilled and one makes contact with the superconsciousness, it begins to show us how our affairs will work out or how we can help to bring about the desired prosperity.

This is the law of mind. The principle is within each one of us, but we must be spiritually quickened in life and in understanding before we can successfully work in accord with it. However we must not discount the understanding of the natural man. The mind in us that reasons and looks to the physical side of things has also the ability to look within. It is the door through which divine ideas must come. Jesus, the Son of man, called Himself "the door" and "the way." It is the divine plan that all expression or demonstration shall come through this gateway of man's mind. But above all this are the ideas that exist in the primal state of Being, and this is the truth of which we must become conscious. We must become aware of the source of our substance. Then we can diminish or increase the appearance of our supply or our finances, for their appearance depends entirely on our understanding and handling of the ideas of substance.

The time is coming when we shall not have to work for things, for our physical needs in the way of food and clothing, because they will come to us through the accumulation of the right ideas in our mind. We will begin to understand that clothing represents one idea of substance, food another, and that every manifest thing is representative of an idea.

In the 2d chapter of Genesis this living substance is called "dust of the ground" in the Hebrew, and Adam was formed from it. We find that the elemental substance is in our body. The kingdom of the heavens or the kingdom of God is within man. It is a kingdom of substance and of Mind. This Mind interpenetrates our mind and our mind interpenetrates and pervades our body. Its substance pervades every atom of our body. Are you giving it your attention, or do you still look to outer sources for supply? Are you meditating and praying for an understanding of this omnipresent substance? If you are, it will come, and it will demonstrate prosperity for you. When it does, you are secure, for nothing can take that true prosperity from you. It is the law that does not and cannot fail to operate when once set in operation in the right way.

This law of prosperity has been proved time and time again. All men who have prospered have used the law, for there is no other way. Perhaps they were not conscious of following definite spiritual methods, yet they have in some way set the law in operation and reaped the benefit of its unfailing action. Others have had to struggle to accomplish the same things. Remember that Elijah had to keep praying and affirming for a long time before he demonstrated the rain. He sent his servant out the first time, and there was no sign of a cloud. He prayed and sent him out again and again with the same result, but at last, after repeated efforts, the servant said he saw a little cloud. Then Elijah told them to prepare for rain, and the rain came. This shows a continuity of effort that is sometimes necessary. If your prosperity does not become manifest as soon as you pray and affirm God as your substance, your supply, and your support, refuse to give up. Show your faith by keeping up the work. You have plenty of Scripture to back you up. Jesus taught it from the beginning to the end of His ministry and demonstrated it on many occasions. Many have done the same thing in His name.

Jesus called the attention of His followers to the inner realm of mind, the kingdom of God substance. He pointed out that the lilies of the field were gloriously clothed, even finer than Solomon in all his glory. We do not have to work laboriously in the outer to accomplish what the lily does so silently and beautifully. Most of us rush around trying to work out our problems for ourselves and in our own way, with one idea, one vision: the material thing we seek. We need to devote more time to silent meditation and like the lilies of the field simply be patient and grow into our demonstrations. We should remember always that these substance ideas with which we are working are eternal ideas that have always existed and will continue to exist, the same ideas that formed this planet in the first place and that sustain it now.

A great German astronomer had worked the greater part of his life with a desire to know more about the stars. One night, quite suddenly and strangely enough--for he had given but little thought to the spiritual side of things--he broke right out into a prayer of thanksgiving because of the perfect order and harmony of the heavens. His prayer was "O God, I am thinking Thy thoughts after Thee." The soul of this man had at that moment made the contact and union with infinite Mind. But though this contact seemed to be made suddenly, it was the result of long study and the preparation of his mind and thought. Jesus expressed the same at-one-ment with God at the moment of His supreme miracle, the raising of Lazarus. His words were "Father, I thank thee that thou heardest me. And I knew that thou hearest me always."

This gives us another side of the prosperity law. We open the way for great demonstrations by recognizing the Presence and praising it, by thanking the Father for Spiritual quickening. We quicken our life by affirming that we are alive with the life of Spirit; our intelligence by affirming our oneness with divine intelligence; and we quicken the indwelling, interpenetrating substance by recognizing and claiming it as our own. We should meditate in this understanding and give sincere thanks to the God of this omnipresent realm of ideas because we can think His thoughts after Him. We can thank the Father that His thoughts are our thoughts and that our natural mind is illumined by Spirit. We can illumine our mind any time by affirming this thought:

I thank Thee, Father, that I think Thy thoughts after Thee and that my understanding is illumined by Spirit.

Spiritual thoughts are infinite in their potentiality, each one being measured by the life, intelligence, and substance with which it is expressed. The thought is brought into expression and activity by the word. Every word is a thought in activity, and when spoken it goes out as a vibratory force that is registered in the all-providing substance.

The mightiest vibration is set up by speaking the name Jesus Christ. This is the name that is named "far above all rule, and authority," the name above all names, holding in itself all power in heaven and in earth. It is the name that has power to mold the universal substance. It is at one with the Father-Mother substance, and when spoken it sets forces into activity that bring results. "Whatsoever ye shall ask of the Father in my name, he may give it to you." "If ye shall ask anything in my name, that will I do." There could be nothing simpler, easier, or freer from conditions in demonstrating supply. "Hitherto [before the name Jesus Christ was given to the world] have ye asked nothing in my name: ask, and ye shall receive, that your joy may be made full."

The sayings of Jesus were of tremendous power because of His consciousness of God. They raised the God ideal far above what had ever before been conceived. These ideas so far transcended the thought plane of the people that even some of the disciples of Jesus would not accept them, and they "walked no more with him." Until fairly recent times most men have failed to grasp the lesson of the power of the spoken word expressing spiritual ideas. Jesus has never been taken literally, else men would have sought to overcome death by keeping His sayings. Few have taken His words in full faith, not only believing them but so saturating their minds with them that they become flesh of their flesh and bone of their bone, being incarnated in their very bodies, as Jesus intended.

The secret of demonstration is to conceive what is true in Being and to carry out the concept in thought, word, and act. If I can conceive a truth, there must be a way by which I can make that truth apparent. If I can conceive of an inexhaustible supply existing in the omnipresent ethers, then there is a way by which I can make that supply manifest. Once your mind accepts this as an axiomatic truth it has arrived at the place where the question of processes begins to be considered.

No one ever fully sees the steps that he must take in reaching a certain end. He may see in a general way that he must proceed from one point to another, but all the details are not definite unless he has gone over the same ground before. So in the demonstration of spiritual powers as they are expressed through man, we must be willing to follow the directions of someone who has proved his understanding of the law by his demonstrations.

We all know intuitively that there is something wrong in a world where poverty prevails and we would not knowingly create a world in which a condition of poverty exists. Lack of any kind is not possible in all God's universe. So when there is an appearance of poverty anywhere, it is our duty to deny it. Sorrow and suffering accompany poverty, and we wish to see them all blotted out. This desire is an index pointing the way to their disappearance. As the consciousness of the kingdom of heaven with its abundant life and substance becomes more and more common among men, these negative conditions will fade out of seeming existence.

Jesus said that all things should be added to those who seek the kingdom of heaven. We do not have to wait until we have fully entered the kingdom or attained a complete understanding of Spirit before prosperity begins to be manifest, but we do have to seek, to turn the attention in that direction. Then things begin to be added unto us. Thousands of people are proving the law in this age. They accept the promise of the Scriptures and are looking to God to supply their every need. In the beginning of their seeking they may have little to encourage them to believe that they will be provided for or helped along any particular line. But they carry out the command to seek and in faith act just as though they were receiving, and gradually there opens up to them new ways of making a living. Sometimes avenues are opened to them to which they are strangers, but they find pleasant experience and are encouraged to continue seeking the kingdom of God and rejoicing in its ever increasing bounty.

Many such people today are wisely using their one talent. They may not have seen the holy of holies in the inner sanctuary, but they are getting closer to it. This is the step we must all take: begin to seek this kingdom of God's substance. Trust in the promise and see the result in the mental currents that are set in motion all about us. You may not be able to see at just what point success began, or what separate word of allegiance to the Father first took effect, but as the weeks or months go by you will observe many changes taking place in your mind, your body, and your affairs. You will find that your ideas have broadened immensely, that your little limited world has been transformed into a big world. You will find your mind more alert and you will see clearly where you were in doubt before, because you have begun thinking about realities instead of appearances. The consciousness of an omnipotent hand guiding all your affairs will establish you in confidence and security, which will extend to the body welfare and surroundings. There will be a lessening or entire absence of prejudice and faultfinding in you. You will be more forgiving and more generous and will not judge harshly. Other people will feel that there has been a change in you and will appreciate you more, showing it in many ways. Things will be coming your way, being added unto you indeed according to the promise.

All this is true not only of your own affairs. The effects extend also to those with whom you come in contact. They will also become more prosperous and happy. They may not in any way connect their improvement with you or your thoughts, but that does not affect the truth about it. All causes are essentially mental, and whoever comes into daily contact with a high order of thinking must take on some of it. Ideas are catching, and no one can live in an atmosphere of true thinking, where high ideas are held, without becoming more or less inoculated with them.

Do not expect miracles to be performed for you, but do expect the law with which you have identified yourself to work out your problem by means of the latent possibilities in and around you. Above all, be yourself. Let the God within you express Himself through you in the world without.

"Ye are gods,

And all of you sons of the Most High."

The idea of God covers a multitude of creative forces. In this case you are working to bring prosperity into your affairs. Hence you should fill your mind with images and thoughts of an all-providing all-supplying Father. The ancient Hebrews understood this. They had seven sacred names for Jehovah, each one of which represented some specific idea of God. They used the name Jehovah-jireh when they wished to concentrate on the aspect of substance. It means "Jehovah will provide," the mighty One whose presence and power provides, regardless of any opposing circumstance. To quicken the consciousness of the presence of God the Hebrews used the name Jehovah-shammah which means "Jehovah is there," "the Lord is present." Realize the Lord present as creative mind, throbbing in the ether as living productiveness.

Charge your mind with statements that express plenty. No particular affirmation will raise anyone from poverty to affluence, yet all affirmations that carry ideas of abundance will lead one into the consciousness that fulfills the law. Deny that lack has any place or reality in your thought or your affairs and affirm plenty as the only appearance. Praise what you have, be it ever so little, and insist that it is constantly growing larger.

Daily concentration of mind on Spirit and its attributes will reveal that the elemental forces that make all material things are here in the ether awaiting our recognition and appropriation. It is not necessary to know all the details of the scientific law in order to demonstrate prosperity. Go into the silence daily at a stated time and concentrate on the substance of Spirit prepared for you from the foundation of the world. This opens up a current of thought that will bring prosperity into your affairs. A good thought to hold in this meditation is this:

The invisible substance is plastic to my abundant thought, and I am rich in mind and in manifestation.

Lesson 3

Faith in the Invisible Substance, the Key to Demonstration

In this lesson we are considering the subject of faith especially as it applies to the demonstration of prosperity. In this study, as in all others, we must start in the one Mind. God had faith when He imaged man and the universe and through His faith brought all things into being. Man, being like God, must also base his creations on faith as the only foundation. Here then is our starting point in building a prosperity consciousness and making our world as we would have it. We all have faith, for it is innate in every man. Our question is how we may put it to work in our affairs.

Jesus gave us our best understanding of faith when He described Peter as a "rock" and asserted that His church, the ecclesia or "called-out ones," was to be built up with this rock or faith as its sure foundation. In this sense faith represents substance, the underlying, basic principle of all manifestation. "Now faith is assurance of things hoped for, a conviction of things not seen."

It is quite possible to possess a reality that cannot be seen, touched, or comprehended by any of the outer senses. It is faith when we are fully conscious of "things not seen" and have the "assurance of things" not yet manifest. In other words, faith is that consciousness in us of the reality of the invisible substance and of the attributes of mind by which we lay hold of it. We must realize that the mind makes real things. "Just a thought" or "just a mere idea," we sometimes lightly say, little thinking that these thoughts and ideas are the eternal realities from which we build our life and our world.

Faith is the perceiving power of the mind linked with a power to shape substance. You hear of a certain proposition that appeals to you and you say, "I have faith in that proposition." Some man whose character seems right is described to you and you say, "I have faith in that man." What do you mean by having faith? You mean that certain characteristics of men or things appeal to you, and these immediately begin a constructive work in your mind. What is that work? It is the work of making the proposition or man real to your consciousness. The character and attributes of the things in your mind become substantial to you because of your faith. The office of faith is to take abstract ideas and give them definite form in substance. Ideas are abstract and formless to us until they become substance, the substance of faith.

A very important work in soul culture is the establishment of a faith substance. Once we discern this law of soul building by faith, we find the Hebrew Scriptures full of illustrations of it. The 1st chapter of Luke's Gospel tells us how Elisabeth and Zacharias were told by an angel that they would have a son and that his name would be John. Zacharias was burning incense at the altar in the exercise of his duties as a priest. This means that when the mind is looking toward Spirit, even if it be in a blind way, and is seeking spiritual things, it will become spiritualized. The burning of incense typifies spiritualization. Zacharias represents the perceptive and Elisabeth the receptive qualities of the soul. When these two work in conjunction in prayer, meditation, and aspiration, the soul is open to the higher thoughts or angels that bring the promise of a new and definite state of consciousness. Zacharias doubted the promise of a son because his wife was past the age of childbearing, and because of his doubts he was stricken dumb. This means that when we perceive spiritual Truth and doubt it, we retard its outer expression; it cannot be spoken into manifestation through us because of our doubt. All the growth is then thrown upon the soul. Elisabeth "hid herself five months," but when the soul begins to feel the presence of the new ego or new state of consciousness, then we again come into faith expression: the speech of Zacharias is restored.

It was the same way in the bringing forth of Jesus. A promise was first made to Mary, and Joseph was assured that the child was the offspring of the Holy Spirit. This represents a still higher step in the work of faith. The bringing forth of John the Baptist is the intellectual perception of Truth. The intellect grasps Truth first. The next step is the bringing forth of substance and life in the subconsciousness. When we have given ourselves entirely to Spirit, we may do things without knowing exactly why. That is because faith is at work in us, and even if we do not know the law and cannot explain faith to the outer consciousness, it continues to do its perfect work and eventually brings forth the demonstration.

Do not fear the power that works out things in the invisible. When you get a strong perception of something that your inner mind tells you is true and good, act on it and your demonstration will come. That is the way a living faith works, and it is the law of your creative word.

Faith can also have understanding added to it. We call our spiritual faculties out of our subconsciousness. When Jesus did some of His most remarkable works He had with Him Peter, James, and John; Peter represents faith, James wisdom or judgment, and John love. These three faculties when expressed together in mind accomplish apparent miracles. You have called out faith in things spiritual, you have faith in God, and you have cultivated your unity with the one Mind; if you then use spiritual judgment and do your work in love, you have become "a teacher in Israel."

In order to have understanding of the law through which we gain or lose in the use of the invisible substance, we must use discrimination or judgment. There is a guiding intelligence always present that we can lay hold of and make our own. It is ours. It belongs to us and it is our birthright both to know it and to use it. Some metaphysicians mistakenly think that they must have hard experiences in order to appreciate the better things of life. They think poverty is a blessing because it educates people to the appreciation of plenty when they get it. They say that it is God's will for us to have some hard times and some good times, feasts and famines. This is not logically true when you consider God as principle. If you think of God as a man who arbitrarily gives or withholds by the exercise of His personal will, you might reach such a conclusion. But God is changeless, and if He gives one moment He will continue to give eternally. It is His nature to give, and His nature is eternally the same. When you talk of hard times, famines, lack, you are talking of something that has no place in the Mind of God. You are not acknowledging God in all your ways but are acknowledging error and affirming that the world has its source in outer things. You must turn around and get into this consciousness, that in Mind, in Spirit, there is abundance.

We often wonder how Jesus could multiply the five loaves and two fishes to meet the hunger of five thousand persons. It was done through a thorough understanding of this law. The five loaves represent the five-sense application of divine substance. The two fishes represent the yeast or multiplying power put into the substance, the source of the increase. We are told that if the yeast of a single setting of bread were allowed to increase, it would fill a space larger than this planet. This shows that there is no limit to the increasing power of elemental substance. It is for us to use as Jesus used this power. It was not a miracle but something that we all have within us as an unawakened ability and that we can learn to develop and use as Jesus did.

Jesus entered into the silence; prayed and blessed the substance at hand. If we would multiply and increase the power, substance, and life in us and at our command, we must get very still and realize that our resource is Spirit, that it is God, and that it is here in all its fullness. We must make contact with it in faith. Then we shall find it welling up within us. Some of you have no doubt had that experience. But if you just let it ooze away without understanding it, you get no benefit. Here is the key to this life and substance you feel when you sit in the silence. You must begin to speak these words with power and authority.

When there is world-wide belief in financial depression, lack of circulation, stagnation, things do not go as we expect and we develop fear, a belief in lack of circulation of money. But if we know the law, we do not come under this fear thought. At any time many persons make money; they use this law and take advantage of opportunity. We should bless everything that we have, for we can increase and multiply what we have by speaking words. Jesus said that His words were spirit and life. Did you ever think that your word is charged with great spiritual life force? It is. Be careful of your words. Man shall be held accountable for his lightest word. If you talk about substance in a negative way, your finances will be decreased, but if you talk about it in an appreciative, large way, you will be prospered.

If we could release the energy in the atoms the scientists tell us about, we could supply the world. This power lies within every one of us. We can begin by freeing the little ideas we have and making them fill the world with thoughts of plenty. We must realize that all power is given to us in heaven and in earth, as Jesus said. He told His apostles that they should receive power when the Holy Spirit had come upon them. They were told to go up into that upper room, in the crown of the head, where spiritual forces begin the formation of new ideas. After you get into the spiritual consciousness and receive the quickening, speak the word with authority and power, concentrating the attention at the power center in the throat. We find it effective to speak the words aloud and then sink back to "the other side" (Galilee), as Jesus often did, to rest and speak them again silently. You can send forth this vibratory energy of Spirit and break down the inertia caused by thoughts of fear and lack, carve out ways, open new avenues to the demonstration of your good.

To bring forth these undeveloped spiritual qualities we must believe in them. "For he that cometh to God must believe that he is." Lord, keep us from unbelief, from leaning on the things we see, from judging according to appearances.

You can conjure up in your mind a thousand imaginary things that will seem real to you. This shows that the mind creates by forming things according to its ideas. The world is awakening in a wonderful way to the truth about the creative power of the mind. Everywhere people are studying psychology or soul culture. The imagination builds things out of the one substance. If you will associate faith with it in its creative work, the things you make will be just as real as those that God makes. Whatever you make in mind and really put faith in will become substantial. Then you must be constantly on your guard as to what you believe, in order that you may bring what is for your good into manifestation.

In what do you have faith? In outer things? If so, you are building shadows without substance, shadows that cease as soon as your supporting thought is withdrawn from them, forms that will pass away and leave you nothing. If you would demonstrate true prosperity, you must turn from things and, as Jesus told His disciples, "have faith in God." Do not have faith in anything less than God, in anything other than the one Mind, for when your faith is centered there, you are building for eternity. Mind and the ideas of Mind will never pass away. There will never be an end to God. There will never be an end to Truth, which God is. There will never be an end to substance, which God is. Build with the divine substance, cultivate faith in realities and "lay up for yourselves treasures in heaven."

The foundation of every work is an idea. Faith is that quality of mind which makes the idea stand out as real, not only to ourselves but to others. When others have faith in the thing you are doing, making, or selling, they see it as real and worth while. Then your success and your prosperity are assured. Only that exists in whose becoming really visible or valuable you have great faith. If you say and believe, "I have faith in the substance of God working in and through me to increase and bring abundance into my world," your faith will start to work mightily in the mind substance and make you prosperous. Whatever you put into substance along with faith will work out in manifestation in your world. We have seen it done and we have proved the law too many times to have any doubt.

The Scriptures are filled with illustrations of this activity of bringing things to pass through faith in substance. The characters of whom we read in the Scriptures represent ideas carrying forward their work in human souls. If we think that they existed only as people of thousands of years ago, we put our faith back thousands of years, instead of letting it work for us this minute in our everyday affairs of life. To demonstrate as Jesus did we must put our faith in the one substance and say, "I have faith in God."

You demonstrate prosperity by an understanding of the prosperity law and by having faith in it, not by appealing to the sympathy of others, trying to get them to do something for you or give you something. Faithfulness and earnestness in the application of the prosperity law will assure you of success.

"Every good gift and every perfect gift is from above, coming down from the Father of lights, with whom can be no variation, neither shadow that is cast by turning."

"In all thy ways acknowledge him,

And he will direct thy paths."

Let us all know that just now we are in the very presence of creative Mind, the Mind that made the universe and everything in it. This Mind is here and at work right now as much as it ever was or ever will be. When we fully realize this, we increase the activity of Mind in us immeasurably. You must realize that God is Spirit and that Spirit is very real and powerful, and by far the most substantial thing in all the world.

It may be hard for those who have become attached to material things to realize that there is an invisible real life and substance that is much more substantial and real than the material. The men of science tell us that the invisible forces have a power that is millions of times more real and substantial than all the material world. When we read statements about some of the recent discoveries of science, which everyone accepts and talks about, we are truly amazed. Such statements made by religionists would be called preposterous and unbelievable. Yet religion has been making the same statements in different ways for thousands of years. Now science is helping religion by proving them.

In comparing substance and matter as regards their relative reality one scientific writer says that matter is merely a crack in the universal substance. It is universal substance that man is handling all the time with his spiritual mind. Through your thoughts you deal with the wonderful spiritual substance, and it takes form in your consciousness according to your thought about it. That is why we must hold the thought of divine wisdom and understanding: so that we may use these creative mind powers righteously. We use them all the time either consciously or unconsciously and we should use them to our advantage and blessing.

Every time you say, "I am a little short of funds," "I haven't as much money as I need," you are putting a limit on the substance in your own consciousness. Is that wisdom? You want a larger supply, not a limited supply of substance. Therefore it is important to watch your thoughts so that the larger supply may come through your mind and into your affairs. Say to yourself, "I am God's offspring, and I must think as God thinks. Therefore I cannot think of any lack or limitation." It is impossible that in this universal Mind that fills everything there can be any such thing as absence. There is no lack of anything anywhere in reality. The only lack is the fear of lack in the mind of man. We do not need to overcome any lack, but we must overcome the fear of lack.

This fear of lack led men to speculate in order to accumulate substance and have a lot of it stored up. This caused a still greater fear of lack in other men, and the situation grew worse and worse until it became generally believed that we must pile up the material symbols of substance for a possible lack in the future. We have tried that system and found that it fails us every time. We must learn to understand the divine law of supply and the original plan, which is that we have each day our daily bread. That is all we really want, just the amount of things we need for today's use, plus the absolute assurance that the supply for tomorrow's needs will be there when tomorrow comes. This assurance cannot be found in hoarding or piling up, as we have learned by experience. It can be had if we have faith and understand the truth about omnipresent, always available substance. Anything less than today's needs is not enough. Anything more than we need for today is a burden. Let us start with the fundamental proposition that there is plenty for you and for me and that the substance is here all the time, supplying us with every needful thing, according to our thought and word.

In the morning, immediately upon awakening, take a quiet meditative thought. A good foundation statement to hold in the silence is:

"Let the words of my mouth and the meditation of my heart

Be acceptable in Thy sight,

O Jehovah, my rock, and my redeemer."

Think of the meaning of these words as you meditate on them. The words of your mouth and the thoughts of your heart are now and always molding the spiritual substance and bringing it into manifestation. They will not be acceptable to the Lord unless they bring into manifestation things that are true, lovely, and altogether good. After your morning meditation, when you have declared the omnipresence and the allness of the good, receive it as true and go forth to the day's activities with faith that all things needful are provided and your good must come. The soil and substance omnipresent has many names.

Jesus called it the kingdom of the heavens. Moses in Genesis named it the Garden of Eden. Science says it is the ether. We live in it as fishes live in the sea, and it lives in us and supplies us with all things according to our thoughts. When you start to your work, pause a moment and declare: "I set God before me this day, to guide and guard, to protect and prosper me." Or: "The Spirit of the Lord goes before me this day and makes my way successful and prosperous." Make this your proclamation for the day. Decree it to be so, and the Lord will bring it to pass. During the day, if a thought of lack or limitation should for a moment disturb you, banish it at once with the statement: "Jehovah is my shepherd; I shall not want."

When your mind comes around again to the subject of prosperity, realize most strongly that your prosperity comes from God. It came with you from God, from your contact with God-Mind in your silence, and your prosperity is right with you wherever you are. Supply may seem to come through outer channels, but your real success depends on your inner hold on the prosperity realization. Be thankful for supply that comes through outer channels, but do not limit God's giving to any one channel. Look unto Him and be prospered.

Some Prosperity Prayers

I am always provided for because I have faith in Thee as my omnipresent abundance.

I have faith in Thee as my almighty resource and I trust Thee to preserve me in my prosperity.

I trust the universal Spirit of prosperity in all my affairs. I come to God because I believe that He is and that He is a rewarder of them that seek alter Him.

Lesson 4
Man, the Inlet and Outlet of Divine Mind

The possessions of the Father are not in stocks and bonds but in the divine possibilities implanted in the mind and soul of every man. Through the mind of man ideas are brought into being. Through the soul of man God's wealth of love finds its expression.

It is well said that the mind is the crucible in which the ideal is transmuted into the real. This process of transformation is the spiritual chemistry we must learn before we are ready to work intelligently in the great laboratory of the Father's substance. There is no lack of material there to form what we will, and we can all draw on it as a resource according to our purpose. Wealth of consciousness will express itself in wealth of manifestation.

One who knows Principle has a certain inner security given him by the understanding of God-Mind. Our affirmations are for the purpose of establishing in our consciousness a broad understanding of the principles on which all life and existence depend. Our religion is based on a science in which ideas are related to Principle and to other ideas in a great universal Mind that works under mental laws. It is not a new religion nor a religious fad but points out the real and the true in any religion. If you know Principle, you are able to know at once whether a religion is founded on facts or has a basis of man-made ideas.

In order to demonstrate Principle we must keep establishing ourselves in certain statements of the law. The more often you present to your mind a proposition that is logical and true the stronger becomes that inner feeling of security to you. The mind of man is built on Truth and the clearer your understanding of Truth is the more substantial your mind becomes. There is a definite and intimate relation between what we call Truth and this universal substance of Being. When the one Mind is called into action in your mind by your thinking about it, it lays hold of the substance by the law of attraction or sympathy of thought. Thus the more you know about God the more successful you will be in handling your body and all your affairs. The more you know about God the healthier you will be, and of course the healthier you are the happier, more beautiful, and better you will be in every way. If you know how to take hold of the universal substance and mold it to your uses, you will be prosperous. Mind substance enters into every little detail of your daily life whether you realize the Truth or not. However, to establish yourself in a certain security in the possession and use of universal life, love, intelligence, and substance, you must get a consciousness of it by first mentally seeing the Truth.

All true action is governed by law. Nothing just happens. There are no miracles. There is no such thing as luck. Nothing comes by chance. All happenings are the result of cause and can be explained under the law of cause and effect. This is a teaching that appeals to the innate logic of our mind, yet we sometimes feel like doubting it when we see things happen that have no apparent cause. These happenings that seem miraculous are controlled by laws that we have not yet learned and result from causes that we have not been able to understand. Man does not demonstrate according to the law but according to his knowledge of the law, and that is why we must seek to learn more of it. God is law and God is changeless. If we would bring forth the perfect creation, we must conform to law and unfold in our mind, body, and affairs as a flower unfolds by the principle of innate life, intelligence, and substance.

The United States Congress establishes laws that rule the acts of all American citizens. Those who keep the laws are rewarded by the protection of the law. Congress does not see to it that men obey the laws. That is left to the executive department of the government. The same thing is true of the universal law. God has ordained the law but does not compel us to follow it. We have free will, and the manner of our doing is left entirely to us. When we know the law and work with it, we are rewarded by its protection and use it to our good. If we break the universal law, we suffer limitations, just as a convicted lawbreaker is limited to a cell or prison. The Holy Spirit is the executive official through whom Divine Mind enforces its laws.

You can see from this consideration that God has bestowed the power of Divine Mind on every man. You are using your organism, body, mind, and soul, to carry out a law that God established as a guide for all creation. If you righteously fulfill this mission, you cannot fail to get the righteous results. If you fail to live in accordance with the law--well, that is your affair. God cannot help it if you are not following the law and by it demonstrating health, happiness, prosperity, and all good. Blackstone said that law is a rule of action. So with God's law: if you follow the rules of action, you will demonstrate Truth. You will have all that God has prepared for you from the foundation of the world.

What are the rules of the law? First, God is good and all His creations are good. When you get that firmly fixed in your mind, you are bound to demonstrate good and nothing but good can come into your world. If you let in the thought that there is such a thing as evil and that you are as liable to evil as to good, then you may have conditions that conform to your idea of evil. But remember, evil and evil conditions are not recognized by Divine Mind. If you have thought of evil as a reality or as having any power over you, change your thought at once and begin to build up good brain cells that never heard about anything but good. Pray thus: I am a child of the absolute good. God is good, and I am good. Everything that comes into my life is good, and I am going to have only the good. Establish this consciousness and only the good will be attracted to you and your life will be a perpetual joy. I cannot tell you why this is true but I know that it is and that you can prove it for yourself to your satisfaction.

If you will start right now with the idea of universal and eternal goodness uppermost in your mind, talk only about the good, and see with the mind's eye everything and everybody as good, then you will soon be demonstrating all kinds of good. Good thoughts will become a habit, and good will manifest itself to you. You will see it everywhere. And people will be saying of you, "I know that that man is good and true. I have confidence in him. He makes me feel the innate goodness of all men." That is the way in which the one Mind expresses itself through man. It is the law. Those who live in accordance with the law will get the desired results. Those who fail to do so will get the opposite results.

The law also applies to our demonstrations of prosperity. We cannot be very happy if we are poor, and nobody needs to be poor. It is a sin to be poor. You may ask whether Jesus cited any example of poverty's being a sin? Yes. You will find it in the story of the prodigal son. That is often used as a text to preach to moral sinners, but a close study of it shows that Jesus was teaching the sin of lack and how to gain plenty. It is a wonderful prosperity lesson.

The prodigal son took his inheritance and went into a far country, where he spent it in riotous living and came to want. When he returned to his father's house he was not accused of moral shortcoming, as we should expect. Instead the father said, "Bring forth quickly the best robe and put it on him." That was a lesson in good apparel. It is a sin to wear poor clothes. This may seem to some to be rather a sordid way of looking at the teaching of Jesus, but we must be honest. We must interpret it as He gave it, not as we think it ought to be.

The next act of the father was to put a gold ring on the prodigal's finger, another evidence of prosperity. The Father's desire for us is unlimited good, not merely the means of a meager existence. The ring symbolizes the unlimited, that to which there is no end. It also represents omnipresence and omnipotence in the manifest world.

When the father gave that ring to the son, he gave him the key to all life activity. It was the symbol of his being a son and heir to all that the father had. "All that is mine is thine." The Father gives us all that He has and is, omnipotence, omniscience, all love, and all substance when we return to the consciousness of His house of plenty.

"Put ... shoes on his feet" was the father's next command to the servants. Feet represent that part of our understanding that comes into contact with earthly conditions. In the head or "upper room" we have the understanding that contacts spiritual conditions, but when we read in Scripture anything about the feet, we may know that it refers to our understanding of things of the material world.

The next thing the father did for his returned son was to proclaim a feast for him. That is not the way we treat moral sinners. We decree punishment for them; we send them to jail. But the Father gives a feast to those who come to Him for supply. He does not dole out only a necessary ration but serves the "fatted calf," universal substance and life in its fullness and richness.

The parable is a great lesson on prosperity, for it shows us that people who are dissipating their substance in sense ways are sinners and eventually fall into a consciousness of lack. It also proves that they may become lawful and prosperous again by returning to the Father-Mind. When there are so many lessons in the Bible for moral delinquents, there is no need to twist the meaning of this parable to that purpose. It is so plainly a lesson on the cause of lack and want. Jesus expressly states that the youth wasted his substance in a "far country," a place where the divine law of plenty was not realized. There is a very close relation between riotous living and want. Persons who waste their substance in sensation come to want in both physical and financial ways. If we would make the right use of the divine substance and the divine law, we must come back to the consciousness of the Father and conserve our body substance. Then health and prosperity will become naturally manifest. If we are not resourceful or secure in our use of the one divine substance, we are not secure in anything. Substance is a very important thing in our world, in fact the foundation of it. Therefore we should be secure in our understanding of it and use it according to God's law.

Then let us enter into the very Truth of Being and observe the divine law. Let us realize that our Father is always here and that we are in a "far country" only when we forget His presence. He is constantly giving us just what we will acknowledge and accept under His law. We can take our inheritance and divorce ourselves in consciousness from the Father, but we shall suffer the results, for then we shall not do things in divine wisdom and divine order, and there will be a "famine" in that land. Let us rather seek the divine wisdom to know how to handle our substance and the law of prosperity will be revealed to us. To come into this realization, declare with faith and all assurance: The all-providing Mind is my resource, and I am secure in my prosperity.

Primitive men did not contend for the products of nature so long as they could easily pick the fruits from the trees and sleep beneath the branches. When they began to live in caves contention arose over the best places, and the strongest were usually the victors. "Success leads to success." Those who were able to take the best did so and proved the law that "whosoever hath, to him shall be given, and he shall have abundance." This seems at first thought to be an unjust law, but it has always prevailed in the affairs of the world. Jesus, the greatest of metaphysicians, taught it as a divine law and gave it His commendation. He could not have done otherwise, for it is a righteous law that man shall have what he earns, that industry, effort, and ability be rewarded and laziness discouraged.

This law operates in every department of being. Those who seek the things that the material realm has to offer usually find them. Those who strive for moral excellence usually attain that goal. Those who aspire to spiritual rewards are also rewarded. The law is that we get what we want and work for, and all experience and history have proved it a good law. If this law were removed, world progress would cease and the race become extinct. Where there is no reward for effort, there will be no effort and society will degenerate. We may talk wisely about the inner urge, but when it has no outer field of action it eventually becomes discouraged and ceases to act.

When men evolve spiritually to a certain degree, they open up inner faculties that connect them with cosmic Mind, and attain results that are sometimes so startling that they seem to be miracle workers. What seems miraculous is the action of forces on planes of consciousness not previously understood. When a man releases the powers of his soul, he does marvels in the sight of the material-minded, but he has not departed from the law. He is merely functioning in a consciousness that has been sporadically manifested by great men in all ages. Man is greater than all the other creations of God-Mind because he has the ability to perceive and to lay hold of the ideas inherent in God-Mind and through faith bring them into manifestation. Thus evolution proceeds by man's laying hold of primal spiritual ideas and expressing them in and through his consciousness.

In the exercise of his I AM identity man needs to develop certain stabilizing ideas. One of them is continuity or loyalty to Truth. In the Scriptures and in life we have many examples of how love sticks to the thing on which it has set its mind. Nothing so tends to stabilize and unify all the other faculties of mind as love. That is why Jesus gave as the greatest commandment that we love God.

When you first begin to think of God as everywhere present substance, your mind will not adhere continuously to the idea. You will drop your attention after a while and think, "I haven't enough to meet all our bills." There you have made a break and have lost momentum in your ongoing, and you must patch it up quickly. Affirm, "I am not going to be led astray. The old ideas are error and they are nothing. They have no power over me. I am going to stick to this proposition. God is love, the substance of my supply."

Ruth, the Moabitish woman, became so attached to Naomi (spiritual thought) that she would not leave her but accompanied her back to Palestine. She was loyal and steadfast because of her love. What was the result of her stick-to-itiveness? She was at first a gleaner; then became the wife of a very rich man and was immortalized as one of the ancestresses of David. This lesson of abiding in our highest ideals is one that we must understand. Nothing is so important as sticking to the ideal and never giving up the work we have set out to accomplish. Affirm the law continuously and be loyal to it and you will become successful in its demonstration.

You have doubtless found that there is a spiritual law that brings into manifestation the thoughts we concentrate our attention on, a divine universal law of mind activity that is unfailing. Some adverse condition of your own thought has prevented a full demonstration. Do not let this swerve you from your loyalty to the law. You may seem to attain results very slowly, but that is the best reason for sticking closely to your ideal and not changing your mind. Be loyal to Principle and the adverse condition will break up. Then the true light will come and the invisible substance you have been faithfully affirming will begin to reveal itself to you in all its fullness of good.

Jesus stressed the idea that God has made abundant provision for all His children, even to the birds of the air and the lilies of the field. The Lord has clothed you with soul substance as gloriously as He did Solomon. But you must have faith in this all-providing substance of good and by your continuity of imagination set it to forming the things you desire. If you are persistent in working this idea in your conscious mind, it will eventually drop down into your subconscious mind and continue to work there where things take form and become manifest. Invisible substance, when your subconsciousness becomes filled with it to the overflowing point, will ooze out, as it were, into all your affairs. You will become more prosperous and successful so gradually, simply, and naturally that you will not realize that it derives from a divine source and in answer to your prayers. We must realize all the while however that whatever we put as seed into the subconscious soil will eventually bring forth after its kind and we must exercise the greatest caution so that we do not think or talk about insufficiency or allow others to talk to us about it. As we sow in mind so shall we reap in manifestation.

Some of our well-meaning friends have a way of loading us up with "hard-times" ideas that disperse this prosperity substance that we have accumulated. Sometimes even one adverse thought will cause it to escape; then we must go back and patch up the broken reservoir of substance thinking. We have to hold it in our mind in all its fullness and we should not let go of it for a minute lest the work of demonstration be delayed. When you retire at night, let your last thought be about the abundance of spiritual substance. See it filling all the house and the minds of all the people in the house. That potent thought will then sink into your subconsciousness and continue to work whether you are asleep or awake.

The law of supply is a divine law. This means that it is a law of mind and must work through mind. God will not go to the grocery and bring food to your table. But when you continue to think about God as your real supply, everything in your mind begins to awaken and to contact the divine substance, and as you mold it in your consciousness, ideas begin to come which will connect you with the visible manifestation. You first get the ideas in consciousness direct from their divine source, and then you begin to demonstrate in the outer. It is an exact law and it is scientific and unfailing. "First the blade, then the ear, then the full grain in the ear."

When you work in harmony with this universal law, every needed thing is abundantly supplied. Your part is simply to fulfill the law; that is, to keep your mind filled with mind substance, to store up spiritual substance until the mind is filled with it and it cannot help but manifest in your affairs in obedience to the law "Whosoever hath, to him shall be given." But you are not fulfilling the law when you allow poverty-stricken thoughts to dwell in your mind. They draw other like thoughts, and your consciousness will have no room for the truth that prosperity is for you. Poverty or prosperity, it all depends on you. All that the Father has is yours, but you alone are responsible for the relationship of the Father's good to your life. Through conscious recognition of your oneness with the Father and His abundance you draw the living substance into visible supply.

Do not hesitate to think that prosperity is for you. Do not feel unworthy. Banish all thoughts of being a martyr to poverty. No one enjoys poverty, but some people seem to enjoy the sympathy and compassion they can excite because of it. Overcome any leaning in that direction and every belief that you were meant to be poor. No one is ever hopeless

until he is resigned to his imagined fate. Think prosperity, talk prosperity, not in general but in specific terms, not as something for the other fellow but as your very own right. Deny every appearance of failure. Stand by your guns and affirm supply, support, and success in the very face of question and doubt, then give thanks for plenty in all your affairs, knowing for a certainty that your good is now being fulfilled in Spirit, in mind, and in manifestation.

A Prosperity Treatment

Twenty-Third Psalm

(Revised)

The Lord is my banker; my credit is good.

He maketh me to lie down in the consciousness of omnipresent abundance;

He giveth me the key to His strongbox.

He restoreth my faith in His riches;

He guideth me in the paths of prosperity for His name's sake.

Yea, though I walk in the very shadow of debt,

I shall fear no evil, for Thou art with me;

Thy silver and Thy gold, they secure me.

Thou preparest a way for me in the presence of the collector;

Thou fillest my wallet with plenty; my measure runneth over.

Surely goodness and plenty will follow me all the days of my life,

And I shall do business in the name of the Lord forever.

Lesson 5

The Law That Governs the Manifestation of Supply

It is safe to say that all men are striving to fulfill the law of their being, but few have understood the law. The law is one of the most important things we can study, because only as we come to understand it and in proportion as we understand it can we comply with its requirements and demonstrate our divine possibilities through it.

In reading the Scriptures we gradually raise our consciousness of them as mere history and begin to apprehend them as setting forth the principle or law of life. We find the great Bible characters fitting into the pattern of our own consciousness, where they represent ideas. This makes the Bible a divine Book of Life rather than merely the history of a people. The idea of the law is symbolized by Moses. In our individual consciousness he is denial, the negative side of the law that precedes its affirmative expression. Moses gave the law as "Thou shalt not." Jesus represents the law in its affirmative expression "Thou shalt love the Lord thy God."

Moses could not go into the Promised Land, the four-dimensional state of consciousness, for there can be no negation there. Joshua, whose name has the same meaning as that of Jesus, entered the Promised Land and opened the way for the Children of Israel. He represents the first step in mind toward that full consciousness of the omnipresence and omnipotence of God that was attained in Jesus. Moses was the lawgiver, and Jesus was, in His own words, the fulfillment of the law.

We must begin to see this four-dimensional world within, with its innate capacity for all things. Everything is right here, all that ever was or ever could be, simply waiting to be brought forth into manifestation. The Lord has prepared a great feast and invited all of us to it, just as Jesus explained in parable. We have right here within and all around us this substance ready for our appropriation or eating. Eating is the outer symbol of mental appropriation. We begin to break bread by breaking the substance of mind, everywhere abundantly provided.

We have discovered that there is within us a life force that can be quickened into greater activity by thinking. Everyone has at some time demonstrated that he could overcome the negative condition of weakness by holding the thought of strength. Sometimes the strength follows the thought immediately, sometimes the thought must be persistently held for days or weeks. In demonstrating the law of ever-present abundance we should and do expect the same results. If the demonstration seems slow in coming, patience and persistence will win. That may be because the poverty consciousness has a tenacious hold and takes effort to be got rid of.

There is a law that governs the manifestation of supply, and we may learn that law and apply it by mental determination and faith in the logical sequences of spiritual realities. We have thought that the laws of God were mysterious and sacred, far removed from the ordinary individual, and that we had better try first to learn the laws of food, of medicine, of a thousand other secondary things. A strict metaphysician looks on all these temporal laws as secondary to the one law of God. That one law, we are told, is to be written in our heart, our inward parts. Then there is something within us that naturally responds to the law of God. If we accept this as true, that we know the one law by an inner intelligence and that all other laws are secondary to it, we are in a position to get results, to demonstrate prosperity.

In the natural world about us we see that everything is governed by law. We are told that the whole animal kingdom is guided by instinct. Many theories have been advanced to explain instinct in terms of material thought. Some philosophers have stated that it is something handed down from one generation to the next, incorporated in germ cells. Whether this is true or not, there is every evidence that there is a law either in or around the cells that controls their formation and duplicates the pattern laid down ages ago in Mother Eve and Father Adam. This is the law written in our inward parts, which is not a figure of speech but a recognized fact. We must look within for the law and not without. The laws we find in the outer are the secondary laws. The infinite, creative Mind has given to every one of us a key to the workings of this unfailing inner law. It is that everything we touch mentally or physically represents substance and that it is limited only by ourselves in our thought capacity. We cannot ask God for more substance, for the universe is full of it. We can and should ask for understanding to lay hold of it with our mind; that is, for an increase in our capacity. Back of the substance is the substance idea, and man is related to the cause side of this idea through his oneness with God.

You may think that you could live better and do more good if you had lots of money. Things would not be a bit better with you if you had a million dollars, unless you also had the understanding to use it for the good of yourself and others. Would you give a child a million dollars to go buy candy and ice cream for himself? We must evolve with our possessions until we get the ability to handle them. Then the law is fulfilled. The supply unfolds at the same rate as the need or ability to use substance is developed. Let us realize this law of unfolding substance and get busy to fulfill it in ourselves by developing our understanding and appreciation of it. We should pray for just as much each day as we need or can handle. "Give us this day our daily bread" is a prayer that conforms to the divine law and answers itself.

Infinite Mind has a lawful way for providing its children with supply for all their needs. Nothing is left to chance. God feeds the birds of the air and clothes the lilies of the field, and He will feed and clothe us unless we make it impossible by our refusal to accept His bounty. Paul said that the fulfilling of the law is love. That is exactly what we must do, love the Lord and love our neighbor as ourselves, and love our work. The law is there, in our inward parts, in our very heart. We know what to do. We don't have to pray or beg for God to give us anything. All we need do is to meditate quietly and affirm the presence and power of the great Giver of all, and then accept the gifts. To be true to the law is to stop looking to the without and to look within for supply. Looking to the within means fixing the mind on God as an ever-present Spirit that is also substance and power. Wrapped up within each of us is a great richness of thoughts. These thoughts are prisoners in the subconsciousness only waiting to be set free to go to work for us. They are waiting for the coming of the Son of God, who releases the prisoners and sets the captives free. This Son is now seeking expression in you; is you. Release your rich thoughts, set free your innate powers, and take from the rich substance of the Father what you will.

Through faith in the overcoming power of Jesus Christ, the sense mind will be overcome and the spiritual mind brought into control of your life and affairs. The sense mind is filled with lacks and limitations; the spiritual mind knows only limitless abundance.

You are linked with the universal spiritual mind through the Christ Mind. It is through the Christ Mind that all things come to you; it is the channel to the all-mind of the Father. Make the unity of wholeness with the Christ Mind. Hold that you are master with the Master, one with the all-providing substance and that your prosperity runneth over. As you begin this process of unifying yourself consciously with the inner life and substance, it will begin to well up within you and to overflow into your affairs, so that you will be prosperous. Remain true to this inner life no matter what the outer appearance may be, and you cannot help but bring the good things of life into manifestation.

All manifest substance flows from a realm of light waves, according to the findings of modern physical science. James says, "Every good gift and every perfect gift is from above, coming down from the Father of lights." This is an exact statement of a scientific law, even to the use of the plural form of the word "lights," for as science states, one or more light particles, electrons, form the atom that is the basis of all material manifestation. God ideas then are the source of all that appears. Accept this as an absolute truth, an all-productive truth, and consciously connect your mind with the Father-Mind. Then you will begin to realize a never-failing prosperity that comes from Being itself.

The German philosopher and poet Goethe says, "The highest and most excellent thing in man is formless, and we should guard against giving it shape in anything less than noble dress." This is a recognition of the truth that man has the capacity within himself to give form to the formless substance. Jesus expressed the law by saying, "Whatsoever thou shalt bind on earth shall be bound in heaven; and whatsoever thou shalt loose on earth shall be loosed in heaven." This heaven is the realm of pure ideas in Mind. We are constantly incorporating these ideas into our mind and giving them form and shape according to our loyalty to Truth.

To every metaphysician this is a very important and very delicate process, because it is through this that we develop our soul. This soul development is often compared to the development of a photographic plate. The light puts the image on the sensitive plate in the first place, or as James says, it is a gift from "the Father of lights." There is then an image on the plate, but it is invisible and unmanifest until it goes through a developing process. Infinite Mind has imaged all its attributes in the soul of every man. But man must develop this image into the clear picture, and much of that work must be done in the dark with perfect faith in the law of manifestation. The photographer works in the darkroom, putting the plate through many processes. Sometimes the developer may make an error in some of the operations and the plate will come out with an imperfect image. So the human manifestation sometimes seems distorted, but the image of perfection imprinted by creative Mind is there. This perfect image is "Christ in you, the hope of glory."

Our body and affairs are first proofs of the development of the picture, but floating in our mind are the higher ideas, the real image to be developed. Our mind is engaged more or less in a chemical process. It is hard to find a line of demarcation between physical and mental chemistry, for they follow the same law. However what has been imaged can be brought out by the proper method of development. Whatever you image yourself as doing, you can do.

In our human understanding we have divorced this imaging power of the mind from the executive power. Now let us bring them together and unify them, for when imagination and will work together all things are possible to man. The will is symbolized in Scripture by the king.

King Solomon was probably the world's richest man, and in so far as the world is concerned he was a great success. He demonstrated prosperity. He did not ask God for riches. Let us note that carefully. He asked God for wisdom, for ideas. God is mind and His gifts are not material but spiritual, not things but ideas. Solomon asked for and received the ideas and then developed them himself. Because he was wise all the world came to his court seeking wisdom and bringing riches in exchange for it. The King of Tyre brought the material he needed to build the Temple. The Queen of Sheba brought him great quantities of gold. From this we should get our cue: ask God for rich ideas (substance) and then put them to work in our affairs.

Do not hesitate to use the divine ideas that come to you, but do not forget their source or foundation. There are many people who are very active executives. The moment they get an idea they make use of it, but oftentimes they do not get far, because they forget the foundation on which such ideas rest and from which we must start to build. With a foundation of Truth, of spiritual ideas and substance we can build an enduring structure of prosperity. It will not be based on a false premise. It will stand when the rains descend and the floods come and the winds blow and beat upon it. We do not desire prosperity today and poverty tomorrow. We should seek for the steady, day-by-day realization of abundant supply.

Jesus understood and used this law of forming the formless substance by the power of imagination and will. When the woman touched the hem of His garment, some of this substance, of which He was vividly conscious, flowed from Him and healed her. He immediately remarked that someone had touched Him. Many had touched Him in the throng and no substance had left His body from those contacts, but the woman of faith was open to receive the healing substance and consciously appropriate it. This proved her faith, and Jesus told her to be of good cheer, for her faith had made her whole. The same substance was available to others who crowded against Him, but only the one who recognized it and laid hold of it received. Even so you and I shall receive no benefit, although substance is everywhere around us and in us, unless we recognize its presence by faith and lay hold of it by the hem of its garment (outer expression).

Jesus recognized the omnipresence of substance when He laid hold of it to multiply the loaves and the fishes. He dwelt in a consciousness of it at all times. Once He told the apostles when they asked Him to eat, "I have meat to eat that ye know not of." He built this divine substance into His body, cell by cell, replacing the mortal flesh with the spiritual substance, until His whole body was immortalized. He demonstrated it and told us how it was done. He said, "He that believeth on me, the works that I do shall he do also; and greater works than these shall he do." Then why are so many people poor, distressed, ill, or troubled? There is a way, a law, and a wisdom to apply the law, and there is an abundance of substance waiting to be formed by each of us into whatsoever we will, when we apply that law as a son of God.

There is an inherent faculty that instinctively lays hold of what it calls its own. Even little children like to have their own toys and to keep them separate from those of other children. There is nothing to be condemned in this, for it is the natural outworking of a divine law. It proves that we know, somewhere in our deepest being, that we have been provided for from the foundation of the world and are entitled to our own portion without question. The power of the mind to draw to us those things to which we are divinely entitled is a power that can be cultivated and should be.

We are now on the verge of a new state of mind in matters financial. Let us do away with the erroneous idea that men must be poor to be righteous. Money is man's instrument, not his master. Money was made for man, not man for money. Only those who put money above man and give it power in their minds by worshiping it, are the "rich" men to whom Jesus referred in His story about the camel and the needle's eye. It is not money that controls men, but the ideas they have about money. Ideas of poverty are just as powerful to enslave men as are ideas of wealth. Every man should be taught how to handle ideas, rather than money, so that they serve him rather than have dominion over him.

Some physical scientists are telling us that the time is near when men will manufacture from the ether, right at hand, everything that they need or desire. Man will not have to wait for seedtime and harvest when he learns to use the power of his mind. When we have that consciousness in which our ideas are tangible, all our demands will be quickly fulfilled by the higher law. Throw into your ideas all the life and power of your concentrated thought, and they will be clothed with reality.

When Jesus went into the wilderness of His (then) untried mental powers He was tempted to turn stones into bread. We all have had this temptation, and most of us have succumbed to it. We get our bread out of material things (stones) instead of out of the words that proceed from the mouth of God. It is the word, the idea, that feeds the soul of

man. That is admitted. But we must realize that it is the word, the idea, that feeds the body and the affairs of man also, for unless the word is recognized and appropriated, there is a lack of the true substance and there is no satisfaction in the food. Fortunately the "Father knoweth that we have need of all these things," and in His compassion and mercy He feeds us with the substance even while we still try to assimilate the stones. If we would seek first the kingdom of God, the substance, the "things" would be added and we should consciously enjoy the fullness of living, the abundant life of Jesus Christ.

There is a universal law of increase. It is not confined to bank accounts but operates on every plane of manifestation. The conscious co-operation of man is necessary to the fullest results in the working of this law. You must use your talent, whatever it may be, in order to increase it. Have faith in the law. Do not reason too much but forge ahead in faith and boldness. If you let yourself think of any person or any outer condition as hindering your increase, this becomes a hindrance to you, for you have applied the law of increase to it. Fear of it may cause you to become timid and bury your talent, which defeats the law. Keep your eyes on the abundant inner reality and do not let the outer appearance cause you to falter.

Do not give too close study to yourself or your present condition. To dwell in mind upon your seeming limitations only prolongs their stay and makes your progress slow. A child loses sight of everything but his increase in size. The boy sees himself as a larger boy, even as a man. It is the childlike mind that finds the kingdom. Then look ahead to the perfect man you are to be in the Spirit and behold yourself as the beloved son in whom the Father is well pleased.

God gives the increase, we are told in the Scripture. This is to be remembered, for we so often think that increase is the result of our personal efforts. Increase comes by the operation of a universal law, and our part is to keep that law. Use the talent of life, and it will expand wonderfully. You do this by talking about life, praising it, and giving God thanks for it. Act as though you were alive and glad to be alive and you will gain a new realization of life, an increase in life itself.

Never allow yourself to come under the control of the "I can't" man. He believes in limitations, wraps his talent in them, and hides it away in the negative earth, and no increase is possible to him. Be positive in Spirit and you will succeed. All the negative talents that are buried away in the depths of material thought can be resurrected by Spirit and made positive, put to the right use, contributing to the increase of your good. Appetite and passion, which are decreasing and destructive in the material can be made increasing and constructive when directed to the things of Spirit. "Blessed are they that hunger and thirst after righteousness: for they shall be filled."

If there is any lack apparent in man's world it is because the requirements of the law of manifestation have not been met. This law is based on mind and its operation through thoughts and words. The key to the operation of mind is symbolically set forth in the Genesis account of the six days of creation. Man's mind goes through the identical steps in bringing an idea into manifestation. Between the perception of an idea and its manifestation there are six definite, positive movements, followed by a (seventh) "day" of rest, in which the mind relaxes and sees its work in process of fulfillment.

In bringing forth a manifestation of God's abundant supply, take the first step by saying, "Let there be light"; that is, let there be understanding. You must have a clear perception of the principle back of the proposition "God will provide." The one universal, eternal, substance of God, which is the source of all, must be discerned and relied on, while dependence on material things must be eliminated from thought. So long as you depend on money alone you are worshiping a false god and have not discerned the light. You must first enter into the understanding that God, omnipresent, omnipotent, and omniscient, is the source and that you can draw on this source without limit. If you have established that light, you have begun your demonstration and can go to the second step. A "firmament" must be established; that is, a firm place in the mind, a dividing of the true from the apparent. This is done through affirmation. As you affirm God as your supply and support, your words will in due season become substance to you, the substance of faith.

The third step is the forming of this substance into tangibility. "Let the dry land appear." Out of the omnipresent substance your mind forms whatever it wants by the power of imagination. If it is food you need, see yourself as bountifully supplied with food. If you have already taken the other steps, you can picture in mind the things you desire and bring them into your manifest world. If the other steps of understanding and faith have not been taken first, there will of course be no demonstration, for above all the creative law is orderly and works by progressive steps. Many people have tried to demonstrate by visualizing and concentrating and have failed because they have put the third step first. They have not developed understanding or faith. If you work according to the law, conforming to its orderly operation as revealed in the degrees of creation, you cannot fail, because when you have fulfilled the law you have found the kingdom.

Jesus recognized order as a fundamental factor in the law of increase. When He fed the multitude He made them sit down in companies. If you study the story carefully you will see that there was a great deal of preliminary preparation before the demonstration was made. There was a recognition of the seed ideas, the loaves and fishes carried by the small boy. There was a prayer of thanks for that supply and then it was blessed. All this preceded the actual appearing and appropriation of the supply. Every demonstration is based on the same law of increase and goes through the same orderly steps.

Pray, but let your prayer be affirmative, for that is the prayer of faith. A begging prayer filled with ifs is a prayer of doubt. Keep praying until affirmations become a habit of mind. The race thought of lack must be penetrated and so charged with the truth of God's omnipresent abundance that all consciousness of lack and poverty disappears from the face of the earth. The more we trust to the simplicity and infallibility of the law the better will be our individual demonstration and the more we shall contribute to the transformation of the race thought that causes lack and famine. Those who make the greatest spiritual demonstrations are not the wise of the world but the obedient children of the law on the bosom of infinite love.

See what you need as already manifest and as yours. Do not put it off to some uncertain future time. God wants you to have it now. Remember always God's omnipresence, and if doubts come in, do not entertain them. Say: "I trust Omnipotence." "I refuse to be anxious about tomorrow or even the next minute. I know that God does provide for the fulfillment of His divine idea, and I am that divine idea." This divine idea is the son, the perfect man, the Christ, brought forth on the sixth day. If you would have your inheritance, you must not omit this sixth-day realization. God expresses Himself as man and works through man to bring perfection into expression.

To give up all anxiety and trust in the Lord does not mean to sit down and do nothing. "My Father worketh even until now, and I work." We are to work as God works; to work with God, as a son follows the occupation of his father. We are to form what God has created. In the 1st chapter of Genesis we see how the Father works. The various steps in His method are clearly pointed out, and we shall have results only as we faithfully follow them.

Some people think of prosperity as something separate from their spiritual experience, "outside the pale" of religion. They live in two worlds: in one for six days of the week when man runs things, and in the other on the seventh day when God is given a chance to show what He can do. It is personality's demonstration when people find themselves complaining of hard times and depression, but it is not the way to demonstrate God in the fullness of all things. Do all things to the glory of God seven days a week rather than one. Take God into all your affairs. Use this thought in the silence and bring God and His law of prosperity into your affairs: I trust Thy universal law of prosperity in all my affairs.

Lesson 6
Wealth of Mind Expresses Itself in Riches

Prosperity, according to Webster, is an advance or gain in anything good or desirable, successful progress toward, or attainment of a desired object. Prosperity does not mean the same thing to any two persons. To the wage earner an increase of a few dollars in the weekly income may seem like wonderful prosperity, for it means an increase in the comfort and welfare of his family. The man who engages in vast enterprises reckons prosperity in larger terms, and does not consider himself prosperous unless things are coming to him in a big way. Between these extremes are many ideas of prosperity, which shows quite plainly that prosperity is not in the possession of things but in the recognition of supply and in the knowledge of free and open access to an inexhaustible storehouse of all that is good or desirable.

In the great Mind of God there is no thought of lack, and such a thought has no rightful place in your mind. It is your birthright to be prosperous, regardless of who you are or where you may be.

Jesus said to all men, "Seek ye first his kingdom, and his righteousness; and all these things shall be added unto you." This does not mean that if you belong to a certain church you will be prospered, for "righteousness" is not conforming to some particular religious belief but to the law of right thinking, regardless of creed, dogma, or religious form. Get into the prosperity thought and you will demonstrate prosperity. Cultivate the habit of thinking about abundance everywhere present, not only in the forms of imagination but in forms without. Jesus did not make a separation between the two as though they were at enmity. He said, "Render therefore unto Caesar the things that are Caesar's; and unto God the things that are God's." Put things in their right relation, the spiritual first and the material following, each where it belongs, and render to each its own.

Realize first of all that prosperity is not wholly a matter of capital or environment but a condition brought about by certain ideas that have been allowed to rule in the consciousness. When these ideas are changed the conditions are changed in spite of environment and all appearances, which must also change to conform to the new ideas. People who come into riches suddenly without building up a consciousness of prosperity soon part from their money. Those who are born and bred to riches usually have plenty all their life even though they never make the effort to earn a dollar for themselves. This is because the ideas of plenty are so interwoven into their thought atmosphere that they are a very part of themselves. They have the prosperity consciousness, in which there is no idea of any condition under which the necessities of life could be lacking.

We are sometimes asked whether we advocate the accumulation of riches. No. The accumulation of riches, as has been explained, is futile unless it is the outgrowth of a rich consciousness. We advocate the accumulation rather of rich ideas, ideas that are useful, constructive, and of service to the well-being of all mankind. The outer manifestation of riches may follow or it may not, but the supply for every need will be forthcoming because the man of rich ideas has confidence in an all-providing power that never fails. He may not have an extra dollar, but his ideas have merit and he has confidence, a combination that cannot fail to attract the money to carry him forward. This is true riches, not an accumulation of money, but access to an inexhaustible resource that can be drawn on at any time to meet any righteous demand. When a person has this rich consciousness there is no necessity for laying up gold or accumulating stocks and bonds or other property to ensure future supply. Such a one may be most generous with his wealth without fear of depletion, because his rich ideas will keep him in constant touch with abundance. Those who have the thought of accumulating material wealth, a thought that is dominant in the world today, are unbalanced. They have a fear of the loss of riches

that makes their tenure insecure. Their prosperity is based on a wrong idea of the source of riches and eventually means disaster. The sin of riches is not in the possession but in the love of money, a material selfishness that leads to soul starvation.

It is not a crime to be rich nor a virtue to be poor, as certain reformers would have us think. The sin lies in hoarding wealth and keeping it from circulating freely to all who need it. Those who put wealth into useful work that contributes to the welfare of the masses are the salvation of the country. Fortunately, there are many in this country who have the prosperity consciousness. If we were all in a poverty consciousness, famines would be as common here as they are in India or China. Millions in those lands are held in the perpetual thought of poverty and they suffer want in all its forms from the cradle to the grave. The burden of the poverty thought reacts on the earth so that year after year it withholds its products and many people starve.

Universal Mind controls all nature and is in possession of all its products. "The earth is the Lord's, and the fullness thereof" is a great Truth. Puny, personal man uses all his craft to get control of the products of nature but is always defeated in the end. Only the universal man of Spirit is in undisputed possession, and to him the Father says, "All that is mine is thine." Jesus did not have title to a foot of land and evidently had no money, for the apostles carried whatever funds the company had.

He did not even burden Himself with a tub, as did Diogenes, and "had not where to lay his head." Yet He was always provided with entertainment of the best. He took it for granted that whatever He needed was His. The fish carried His pocketbook, and the invisible ethers furnished the sideboard from which He handed out food for thousands. He was rich in every way for He had the prosperity consciousness and proved that the earth with all its fullness does belong to the Lord, whose righteous sons are heirs to and in possession of all things.

The anxious thought must be eliminated and the perfect abandon of the child of nature assumed, and when to this attitude you add the realization of unlimited resources, you have fulfilled the divine law of prosperity.

The imagination is a wonderful creative power. It builds all things out of the one substance. When you associate it with faith, you make things just as real as those that God makes, for man is a co-creator with God. Whatever you form in the mind and have faith in will become substantial. Then you should be on guard as regards what you put your faith in. If it is material forms, shadows that cease to be as soon as your supporting thought is withdrawn from them, you are building temporary substance that will pass away and leave you nothing. Put your faith in the real or, as Jesus told His disciples, "have faith in God."

The real search of all people is for God. They may think they are looking for other things, but they must eventually admit that it is God they seek. Having once felt His presence within them, they are keenly conscious that only God can satisfy. The place where we meet God should be made so sure and so pure that we can never mistake His voice or be hidden from His face. This place we know as the mind, the inmost recess of the soul, the kingdom of the heavens within us.

It is not sufficient however to sit down and hold thoughts of abundance without further effort. That is limiting the law to thought alone, and we want it to be fulfilled in manifestation as well. Cultivating ideas of abundance is the first step in the process. The ideas that come must be used. Be alert in doing whatever comes to you to do, cheerful and competent in the doing, sure of the results, for it is the second step in the fulfilling of the law.

You can do anything with the thoughts of your mind. They are yours and under your control. You can direct them, coerce them, hush them, or crush them. You can dissolve one thought and put another in its stead. There is no other place in the universe where you are the absolute master. The dominion given you as your divine right is over your own thoughts only. When you fully apprehend this and begin to exercise your God-given dominion, you begin to find the way to God, the only door to God, the door of mind and thought.

If you are fearful that you will not be provided with the necessities of life for tomorrow, next week, or next year, or for your old age, or that your children will be left in want, deny the thought. Do not allow yourself for a moment to think of something that must be outside the realm of all-careful, all-providing good. You know even from your outer experience that the universe is self-sustaining and that its equilibrium is established by law. The same law that sustains all sustains you as a part. Claim your identity under that law, your oneness with the all, and rest in the everlasting arms of Cause, which knows nothing of lack. If you are in a condition of poverty, this attitude of mind will attract to you opportunities to better your condition. Insulate your mind from the destructive thoughts of all those who labor under the belief in hard times. If your associates talk about the financial stringency, affirm all the more persistently your dependence on the abundance of God.

By doing this you place yourself under a divine law of demand and supply that is never influenced by the fluctuations of the market or the opinions of men. Every time you send out a thought of wholehearted faith in the I AM part of yourself, you set in motion a chain of causes that must bring the results you seek. Ask whatsoever you will in the name of the Christ, the I AM, the divine within, and your demands will be fulfilled; both heaven and earth will hasten to do your bidding. But when you have asked for something, be on the alert to receive it when it comes. People complain that their prayers are not answered when, if we knew the truth, they are not awake to receive the answer when it comes.

If you ask for money, do not look for an angel from the skies to bring it on a golden platter, but keep your eyes open for some fresh opportunity to make money, an opportunity that will come as sure as you live.

These are some tangible steps along the way to the larger manifestation you desire. No one is ever given the keys to the Father's storehouse of wealth until he has proved his faith and his reliability. Then he may go in and pass out the goods freely. If the men of the world, with their selfish ideas of "mine and thine," were given the power, without a thorough mental cleansing, of instantly producing whatever they desire, they would undoubtedly practice still greater oppressions on their fellows, and existing conditions would not be improved.

A stonecutter sees a block of marble as so many hours work, while Michelangelo sees it as an angel that it is his privilege to bring forth. This is the difference between those who see the material world as so much matter and those who look on it with the eyes of mind and the imagination that works toward perfection. One who paints a picture or makes a piece of sculpture first sees it in his mind. He first imagines or images it. If he wants a strong picture he makes force one of the elements of his image. If he wants beauty and character, he puts love into it. He may not see the perfect picture until all these elements are combined, then it requires but little effort to transfer it from his mind to the canvas or to the marble.

On the sixth day of creation, we are told, God "imaged" His man, made him in His image and likeness. This does not mean that God looks like man, a personal being with manlike form. We make a thing in our own image, the image we have in mind for the thing, and our creation does not resemble us in any way. God is without form, for He is Spirit. God is an idea that man has tried to objectify in various forms. He is the universal substance, the life that animates the substance, and the love that binds it together. Man just naturally gives some form to every idea he has, even the idea of God, for the formative faculty of the mind is always at work whether we are awake or asleep. We get material for forming mental pictures from without and from within.

This imaging or formative power of the mind could not make anything unless it had the substance out of which to form it. One could not make a loaf of bread without the flour and other ingredients. Yet with all the ingredients at hand one could not make a loaf unless one had the power of imaging the loaf in one's mind. This seems simple, but the fact is that the power to form the loaf is less common than the available material for the loaf. Flour and water are abundant, but only certain people can use them in the right way to form a palatable loaf of bread. So with this subject of prosperity. Substance is everywhere, filling all the universe. There is no lack. If we have not been successful in forming it into the things we have needed and wished for, it is not because of lack of substance but of lack of understanding how to use our imaging power.

The world goes through periods of seeming lack because the people have refused to build their prosperity on the inner, omnipresent, enduring substance, and on the contrary have tried to base it on the substance that they see in the outer. This outer substance, formed by the imaging power of men in past ages, seems to be limited, and men struggle for it, forgetting their own divine power to form their own substance from the limitless supply within. The lesson for all of us should be to build our prosperity on the inner substance.

Those who do demonstrate prosperity through the law of men have nothing permanent. All their possessions may be swept away in a moment. They have not built on the orderly law of God, and without the rich thoughts of God's bounteousness no one can have an enduring consciousness of supply.

No disease, poverty, or any other negative condition can enter into our domain unless we invite it. Nor can it remain with us unless we entertain it. Conscious power over all such conditions is one of our greatest delights and a part of our divine inheritance, but we must learn the law and apply the power in the right way.

Men have a consciousness of lack because they let Satan, the serpent of sense, tempt them. The Garden of Eden is within us here and now, and the subtle temptation to eat of the tree of sensation is also still with us. We have been given dominion over the animal forces of the body, the "beasts of the field," and must tame them, making them servants instead of masters of the body. Instead of feeding them we must make them feed us. When we overcome the animals within, it will be easy to train them in the without. This truth of overcoming is taught all through the Scriptures, and we can demonstrate it in our life, for God has endowed us with the power to overcome. We must lay hold of that inherent power and begin to use it constructively.

The whole human family seems to be sensation mad. All our economic and social troubles can be traced right back to the selfishness of the sense man. We can never overcome these conditions in the outer until we overcome their causes in the inner soul of ourselves. There is sure to be repetition of war and peace, plenty and famine, good times and depressions until we take the control of mind substance away from the sense man and give it to the spiritual man. We know that there is a spiritual man and we look forward in some ideal way to his coming, but he will never come until we bring him. We hope and pray for the coming of better things; but as Mark Twain said about the weather, "no one does anything about it." We can do something about the matter of self-control and each of us must if we are ever to improve our condition physically and financially as well as morally and spiritually.

We must lift up this serpent of sense, as Moses lifted up the serpent in the wilderness, and control it in the name of Christ.

Eliminate all negative thoughts that come into your mind. Yet do not spend all your time in denials but give much of it to the clear realization of the everywhere present and waiting substance and life. Some of us have in a measure inherited "hard times" by entertaining the race thought so prevalent around us. Do not allow yourself to do this. Remember your identity, that you are a son of God and that your inheritance is from Him. You are the heir to all that the Father has. Let the I AM save you from every negative thought. The arrows that fly by day and the pestilence that threatens are these negative race thoughts in the mental atmosphere. The I AM consciousness, your Saviour, will lead you out of the desert of negation and into the Promised Land of plenty that flows with milk and honey.

Deny that you can lose anything. Let go of negative thoughts of financial loss or any other kind of loss and realize that nothing is ever lost in all the universe. There are opportunities everywhere, just as there have always been, to produce all that you need financially, or otherwise. God wants you to be a producer of new ideas. New ideas come to you from within. Do not think for a moment that you are limited to the ideas that come from without. Many of those ideas are outgrown anyway and have outlived their usefulness. That is why we go through periods of change; so that old outworn ideas

can be discarded and replaced with new and better ones. There have been more inventions since the beginning of the so-called depression than in any previous similar period of American history. This shows that new ideas are within man, just waiting to be called out and put into expression. We can find new ways of living and new methods of work; we are not confined to the ways and methods of the past. When we commune with the Spirit within and ask for new ideas, they are always forthcoming. When these ideas from within us are recognized, they go to work and come to the surface. Then all the thoughts we have ever had, as well as the thoughts of other people, are added to them and new things are quickly produced. Let us quit slavishly depending on someone else for everything and become producers, for only in that direction lies happiness and success. Let us begin to concentrate on this inner man, this powerful man who produces things, who gets his ideas from a higher-dimensional realm, who brings ideas from a new territory, the land of Canaan.

What kind of character are you giving to this inner substance by your thoughts? Change your thought and increase your substance in the mind, as Elisha increased the oil for the widow. Get larger receptacles and plenty of them. Even a very small idea of substance may be added to and increased. The widow had a very small amount of oil, but as the prophet blessed it it increased until it filled every vessel she could borrow from the neighbors. We should form the habit of blessing everything that we have. It may seem foolish to some persons that we bless our nickels, dimes, and dollars, but we know that we are setting the law of increase into operation. All substance is one and connected, whether in the visible or the invisible. The mind likes something that is already formed and tangible for a suggestion to take hold of. With this image the mind sets to work to draw like substance from the invisible realm and thus increase what we have in hand. Jesus used the small quantity of loaves and fishes to produce a great quantity of--loaves and fishes. Elisha used a small amount of oil to produce a great amount of--oil. So when we bless our money or other goods, we are complying with a divine law of increase that has been demonstrated many times.

Another step in the demonstration of prosperity is the preparation of the consciousness to receive the increase. If we pray for rain, we should be sure that we have our umbrellas with us. You read in the 3d chapter of II Kings how Elisha caused the water to come from the invisible and fill trenches in the desert. But first the trenches had to be dug in the dry ground. That required faith, but the kings had it, and they dug trenches all over a large valley, just as Elisha had commanded. It was through the understanding of Elisha, who knew the truth about the invisible substance, that this seeming miracle was accomplished. Yet the trenches had to be prepared, and you must prepare your consciousness for the inflow of the universal substance.

It obeys the law of nature, just as does water or any other visible thing, and flows into the place prepared for it. It fills everything you hold in your mind, whether vessels, trenches, or your purse.

It is not advisable to hold for too specific a demand. You might visualize a hundred dollars and get it when a thousand was coming your way. Do not limit the substance, to what you think you need or want; rather broaden your consciousness and give infinite Mind freedom to work, and every good and needful thing will be provided you. Make your statements broad and comprehensive so that your mind may expand to the Infinite rather than trying to cram the Infinite into your mind.

Statements To Broaden The Mind And Fill It With The Richness Of Substance

Infinite wisdom guides me, divine love prospers me, and I am successful in everything I undertake.

In quietness and confidence I affirm the drawing power of divine love as my magnet of constantly increasing supply.

I have unbounded faith in the omnipresent substance increasing and multiplying at my word of plenty, plenty, plenty.

Father, I thank Thee for unlimited increase in mind, money, and affairs.

Lesson 7
God Has Provided Prosperity for Every Home

The home is the heart of the nation. The heart is the love center. Love is the world's greatest attractive power. The electromagnet that lifts the ingots of steel must first be charged with the electric current, for without the current it is powerless. So the heart of man, or the home that is the heart of the nation, must be aglow with God's love; then it becomes a magnet drawing all good from every direction. God has amply provided for every home, but the provision is in universal substance, which responds only to law. Through the application of the law the substance is drawn to us and begins to work for us.

It is the law of love that we have whatsoever we desire. As a father gives his children gifts so the Lord gives to us, because of love. When we desire aright, we put our thoughts into the supermind realm; we contact God-Mind and from it draw the invisible substance that is manifest in temporal things. The substance thus becomes a part of our mind and through it of our affairs. We draw spiritual substance to ourselves just as the magnet draws the iron. When we think about the love of God drawing to us the substance necessary for support and supply, that substance begins to accumulate all around us, and as we abide in the consciousness of it, it begins to manifest itself in all our affairs.

"Perfect love casteth out fear." Fear is a great breeder of poverty, for it breaks down positive thoughts. Negative thoughts bring negative conditions in their train. The first thing to do in making a demonstration of prosperity in the home is to discard all negative thoughts and words. Build up a positive thought atmosphere in the home, an atmosphere that is free from fear and filled with love. Do not allow any words of poverty or lack to limit the attractive power of love in the home. Select carefully only those words that charge the home atmosphere with the idea of plenty, for like attracts like in the unseen as well as the seen. Never make an assertion in the home, no matter how true it may look on the surface, that you would not want to see persist in the home. By talking poverty and lack you are making a comfortable place for these unwelcome guests by your fireside, and they will want to stay. Rather fill the home with thoughts and words of plenty, of love, and of God's substance; then the unwelcome guests will soon leave you.

Do not say that money is scarce; the very statement will scare money away from you. Do not say that times are hard with you; the very words will tighten your purse strings until Omnipotence itself cannot slip a dime into it. Begin now to talk plenty, think plenty, and give thanks for plenty. Enlist all the members of the home in the same work. Make it a game. It's lots of fun, and, better than that, it actually works.

Every home can be prosperous, and there should be no poverty-stricken homes, for they are caused only by inharmony, fear, negative thinking and speaking. Every visible item of wealth can be traced to an invisible source. Food comes from grain, which was planted in the earth; but who sees or knows the quickening love that touches the seed and makes it bear a hundredfold? An unseen force from an invisible source acts on the tiny seeds, and supply for the multitude springs forth.

The physical substance that we name earth is the visible form of a superabundant mind substance, everywhere present, pervading all things, and inspiring all things to action. When the grain or seed is put into the earth, the quickening thought of the universe causes the little life germ to lay hold of the spiritual substance all about it and what we call matter proves to be a form of mind. "There is no matter; all is mind."

Words are also seeds, and when dropped into the invisible spiritual substance, they grow and bring forth after their kind. "Do men gather grapes of thorns, or figs of thistles?" Farmers and gardeners choose their seed with the greatest care. They reject every defective seed they find and in this way make sure of the coming crop. To have prosperity in your home you will have to exercise the same intelligent discrimination in the choice of your seed words.

You should expect prosperity when you keep the prosperity law. Therefore, be thankful for every blessing that you gain and as deeply grateful for every demonstration as for an unexpected treasure dropped into your lap. This will keep your heart fresh; for true thanksgiving may be likened to rain falling upon ready soil, refreshing it and increasing its productiveness. When Jesus had only a small supply He gave thanks for the little He had. This increased that little into such an abundance that a multitude was satisfied with food and much was left over. Blessing has not lost its power since the time Jesus used it. Try it and you will prove its efficacy. The same power of multiplication is in it today. Praise and thanksgiving impart the quickening spiritual power that produces growth and increase in all things.

You should never condemn anything in your home. If you want new articles of furniture or new clothes to take the place of those you now have, do not talk about your present things as old or shabby. Watch your words. See yourself clothed as befits a child of the King and see your house furnished just as pleases your ideal. Thus plant in the home atmosphere the seed of richness and abundance. It will all come to you. Use the patience, the wisdom, and the assiduity that the farmer employs in planting and cultivating, and your crop will be sure.

Your words of Truth are energized and vitalized by the living Spirit. Your mind is now open and receptive to an influx of divine ideas that will inspire you with the understanding of the potency of your own thoughts and words. You are prospered. Your home is a magnet of love, drawing to it all good from the unfailing and inexhaustible reservoir of supply. Your increase comes because of your righteous application of God's law in your home.

"The blessing of Jehovah, it maketh rich; And he addeth no sorrow therewith."

Jesus showed men how to live in rest and peace, a simple life. Where the simplicity of His teaching is received and appreciated the people change their manner of living, doing away with ostentation and getting down to the simplicity and beauty of the things that are worth while. Every summer those who feel that they can, plan to go away for a vacation and many of them enjoy a small cabin in the woods where they can live a simple and natural life close to nature. This shows that they long to let go of the burdens of conventionality and rest in touch with the real of things. The soul wearies of the wear and tear of the artificial world, and now and then it must have a season of rest. Jesus invites, "Come unto me, all ye that labor and are heavy laden, and I will give you rest."

There is a great difference between the simple life and poverty. The two have been associated in the minds of some people, and this is the reason they shun the idea of the simple life. Even those who have come into some degree of spiritual understanding sometimes put out of mind all thought of a simple manner of living, because they fear that others will think they are failing to demonstrate prosperity. In such cases those who judge should remember to "judge not according to appearance," and those who are judged should be satisfied with the praise of God rather than with the praise of men. All those who base their prosperity on possessions alone have a purely material prosperity which, though it may seem great for a time, will vanish, because it is founded on the changing of the external and has no root within the consciousness.

There is a great similarity in the homes of nearly all people who have about the same-sized incomes. Each one unconsciously follows suggestion and furnishes his home with the same sort of things as his neighbors. Here and there are exceptions. Someone is expressing his or her individuality, overcoming mass suggestion and buying the kind of furniture he really wants or that is really comfortable and useful. This free, independent spirit has much in its favor in making a prosperity demonstration. The delusion that it is necessary to be just like other people or to have as much as other people have, causes a spirit of anxiety that hinders the exercise of faith in demonstration.

The simple life does not imply poverty and it is not ascetic. It is as different from the austere as it is from wanton luxury. It is the natural, free, childlike, mode of living, and one never really knows what true prosperity is until one comes into this simplicity and independence of spirit. The simple life is a state of consciousness. It is peace, contentment, and satisfaction in the joy of living and loving, and it is attained through thinking about God and worshiping Him in spirit and in truth.

You want to learn how to demonstrate prosperity in your home by the righteous exercise of powers and faculties that God has given you. Realize in the very beginning that you do have these powers and faculties. You are in possession of everything necessary for the demonstration of prosperity and can undertake it with the utmost confidence and faith. You can draw on the omnipresent substance throughout all eternity, yet it will never grow less, for it consists of ideas. Through thinking you take some of these ideas into your mind and they begin to become manifest in your affairs.

Love is one of the ideas that provide a key to the infinite storehouse of abundance. It opens up generosity in us. It opens up generosity in others when we begin to love and bless them. Will it also open up a spirit of generosity in God? It certainly will and does. If you consciously love and bless God, you will soon find that things are coming your way. It will surprise you that just thinking about God will draw to you the things you want and expect, and bring many other blessings that you had not even thought about. Thousands of persons have proved this law to their entire satisfaction, and we have many records that illustrate how people have demonstrated abundance in the very face of apparent lack, simply by thinking about the love of God and thanking Him for what they have. This law will demonstrate itself for you or for anyone who applies it faithfully, for "love never faileth."

Men in business and industry have demonstrated great amounts of money through love. They did not love God, but the love of money attracted the money to them. It drew the substance right to them and enabled them to accumulate money, but merely as material, without the divine idea that assures permanence. We hear about men in high finance going bankrupt quite as often as we hear about men making great fortunes. When we develop a spiritual consciousness, we transfer this personal love to a higher and more stable plane, from the love of money and material things to the love of God, and thus conceived it will attract to us all the resources of infinite Mind forever and ever. Once make a connection with the universal bank of God and you have a permanent source of wealth.

Jesus said that when we come to the altar to make an offering, we should have nothing in our heart against our brother. He said that before we can make contact with the love and power of God we must first make peace with our brother. This means that we must cultivate a love for our fellows in order to set the attractive force of love into operation. All we need do is quicken our love for others by thinking about love and casting out of our mind all hate and fear that would weaken the perfect working of that mighty magnet. As love attracts, hate dissipates. Before you approach God's altar of plenty, go and make

friends with your brother men. Make friends even with the money powers. Do not envy the rich. Never condemn those who have money merely because they have it and you do not. Do not question how they got their money and wonder whether or not they are honest. All that is none of your business. Your business is to get what belongs to you, and you do that by thinking about the omnipresent substance of God and how you can lay hold of it through love. Get in touch with God riches in spirit, lay hold of them by love, and you will have sufficient for every day. "Love therefore is the fulfillment of the law."

The eternal law of Spirit goes right on operating regardless of what you may think, say, or do. It is ordained that love will bring you prosperity, and you need not wonder whether it will or how it will. "Be not therefore anxious, saying, What shall we eat? or, What shall we drink? or, Wherewithal shall we be clothed?" Do not worry. Worry is a thief and a robber, for it keeps your good from you. It breaks the drawing law of love, the law that says, "Perfect love casteth out fear." Banish worry by quietly and confidently affirming the drawing power of divine love as the constantly active magnet that attracts your unfailing supply. A good affirmation to rout worry is one like this:

Divine love bountifully supplies and increases substance to meet my every need.

Nearly all books or articles that deal with success or prosperity stress the well-known virtues of honesty, industry, system and order, faithfulness, hard work. These make an excellent foundation and can be developed. Anyone with determination and will can overcome habits of laziness, carelessness, and weakness. The use of the will is very important in the demonstration of prosperity. If there is disorder or lack of system in your home, overcome it. Affirm: I will to be orderly. I will be orderly. I will be systematic in all my work and affairs. I am systematic. I am orderly. I am efficient.

It takes the use of the will to be persistent, and we must be persistent in making demonstrations. Spasmodic efforts count for little, and many people give up too easily. If things don't come out just right the first time they try, they say the law is wrong and make no further effort. Anything so much worth while as prosperity in the home, and especially a permanent and unfailing supply that continues to meet the daily needs year after year, is worth any effort that we can make. Then be patient but be persistent. Declare: I am not discouraged. I am persistent. I go forward.

When success fails to crown our very first efforts we become discouraged and quit. Then we try to console ourselves with the old thought that it is God's will for us to be poor. Poverty is not God's will, but man lays it to the charge of God to excuse his own feeling of inadequacy and defeat. God's will is health, happiness, and prosperity for every man; and to have all that is good and beautiful in the home is to express God's will for us. God's will is not expressed in a hovel, nor in any home where discord, lack, and unhappiness are entertained. Even a human guest would not stay long in such a home. To have a prosperous home prepare it as the abiding place of God, who gives prosperity to all His children and adds no sorrow therewith. Determine to know God's will and do it. Affirm: I am determined to achieve success through doing God's will. That sums up the whole law. God is more willing to give than we are to receive. What we need to do is to determine what is His will, what He is trying to give, and open ourselves to receive His bounty. We do that by willing to do His will. You can be and have anything that you will to be and to have. Will to be healthy. Will to be happy. Will to be prosperous.

There are many persons who will to be prosperous and who have made up their minds, as they think, very determinedly. But they have not overcome all doubts, and when their demonstration is delayed, as it is in such cases, the doubt increases until they lose faith altogether. What they need is more persistence and determination. The word determined is a good word, a strong, substantial word with power in it. Jesus said that His words were spirit and life and would never pass away. Emerson says that words are alive and if you cut one it will bleed. Use the word determined and emphasize it in your affirmations. If things do not seem to come fast enough, determine that you will be patient. If negative thoughts creep in, determine to be positive. If you feel worried about the results, determine to be optimistic. In response to every thought of lack or need determine to be prosperous. The Lord has prosperity to give, and those who are determined go after their share.

Jesus was quite positive and very determined in all His affirmations. He made big claims for God, and demonstrated them. Without the slightest doubt that the money would be there, He told Peter to put his hand into the fish's mouth and take out the wanted money. His prayers were made of one strong affirmation after another. The Lord's Prayer is a series of determined affirmations. We claim the will of God is for us to be rich, prosperous, and successful. Make up your mind that such is God's will for you and your home and you will make your demonstration.

In the Old Testament, in the 4th chapter of II Kings, there is a fine prosperity lesson for any home. The widow represents one who has lost his consciousness of God's supply and support. That divine idea of God as all-abundance is our true support. The two children of this home represent the thoughts of debt, what the family owes, and what someone owes the family. The prophet is divine understanding. The house is the body consciousness. The pot of oil is faith in spiritual substance. The neighbors are outside thoughts, and their "empty vessels" are thoughts of lack. To go in "and shut the door," as the widow was told to do, is to enter the inner consciousness and shut out the thoughts of lack. This is followed by strong words of affirmation: "pouring" the substance into all the places that seem to be empty or to lack, until all are full. In conclusion it is affirmed that every obligation is met, every debt paid, and there is so much left over that there are no vessels left to hold it.

This compares with the promise of God "I will ... open you the windows of heaven, and pour you out a blessing, that there shall not be room enough to receive it." "Heaven" represents the mind. All this is done in the mind, and you can do it. Carry each step forward in your imagination exactly as if it were occurring in the without. Form your prosperity demonstration in your mind, then hold to the divine law of fulfillment. "And, having done all ... stand." You may not be able to fill all the vessels with oil on your first attempt, but as you practice the method day by day your faith will increase and your results will be in proportion to your increasing faith.

Work at the problem until you prove it. Apply the principle and the solution is sure. If it does not come at once, check over your methods carefully and see wherein your work has not been true. Do not allow one empty thought to exist in your mind but fill every nook and corner of it with the word plenty, plenty, plenty.

If your purse seems empty, deny the lack and say, "You are filled even now, with the bounty of God, my Father, who supplies all my wants." If your rooms are empty, deny the appearance and determine that prosperity is manifest in every part of every room. Never think of yourself as poor or needy. Do not talk about hard times or the necessity for strict economy. Even "the walls have ears" and, unfortunately, memories too. Do not think how little you have but how much you have. Turn the telescope of your imagination around and look through the other end. "Revile not the king, no, not in thy thought; and revile not the rich in thy bedchamber: for a bird of the heavens shall carry the voice, and that which hath wings shall tell the matter."

"Blessed is the man that walketh not in the counsel of the wicked,

Nor standeth in the way of sinners,

Nor sitteth in the seat of scoffers:

But his delight is in the law of Jehovah;

And on his law doth he meditate day and night.

And he shall be like a tree planted by the streams of water.

That bringeth forth its fruit in its season,

Whose leaf also doth not wither;

And whatsoever he doeth shall prosper."

"Through wisdom is a house builded;

And by understanding it is established;

And by knowledge are the chambers filled

With all precious and pleasant riches."

"Jehovah will open unto thee his good treasure."

"And the Almighty will be thy treasure,

And precious silver unto thee."

"Jehovah is my shepherd; I shall not want."

"Trust in Jehovah, and do good;

Dwell in the land, and feed on his faithfulness."

"Jehovah will give grace and glory;

No good thing will he withhold from them that walk uprightly."

"That I may cause those that love me to inherit substance,

And that I may fill their treasuries."

"If ye be willing and obedient, ye shall eat the good of the land."

Lesson 8
God Will Pay Your Debts

Forgive us our debts, as we also have forgiven our debtors." In these words Jesus expressed an infallible law of mind, the law that one idea must be dissolved before another can take its place. If you have in your mind any thought that someone has wronged you, you cannot let in the cleansing power of Spirit and the richness of spiritual substance until you have cast out the thought of the wrong, have forgiven it fully. You may be wondering why you have failed to get spiritual illumination or to find the consciousness of spiritual substance. Perhaps the reason is here: a lack of room for the true thoughts because other thoughts fill your mind. If you are not receiving the spiritual understanding you feel you should have, you should search your mind carefully for unforgiving thoughts. "Thoughts are things" and occupy space in the mind realm. They have substance and form and may easily be taken as permanent by one not endowed with spiritual discernment. They bring forth fruit according to the seed planted in the mind, but they are not enduring unless founded in Spirit. Thoughts are alive and are endowed by the thinker with a secondary thinking power; that is, the thought entity that the I AM forms assumes an ego and begins to think on its own account. Thoughts also think but only with the power you give to them.

Tell me what kind of thoughts you are holding about yourself and your neighbors, and I can tell you just what you may expect in the way of health, finances, and harmony in your home. Are you suspicious of your neighbors? You cannot love and trust in God if you hate and distrust men. The two ideas love and hate, or trust and mistrust, simply cannot both be present in your mind at one time, and when you are entertaining one, you may be sure the other is absent. Trust other people and use the power that you accumulate from that act to trust God. There is magic in it: it works wonders; love and trust are dynamic, vital powers. Are you accusing men of being thieves, and fear that they are going to take away from you something that is your own? With such a thought generating fear and even terror in your mind and filling your consciousness with darkness, where is there room for the Father's light of protection? Rather build walls of love and substance around yourself. Send out swift, invisible messengers of love and trust for your protection. They are better guards than policemen or detectives.

Do not judge others as regards their guilt or innocence. Consider yourself and how you stand in the sight of the Father for having thoughts about another's guilt. Begin your reform with yourself. That means much to one who enjoys an understanding of mind and its laws, though it may mean little to the ordinary individual. He who knows himself superficially, just his external personality, thinks he has reformed when he has conformed to the moral and governmental laws. He may even be filled with his own self-righteousness and daily lift up his voice to praise God that he is not as other men are, that he has forgiven men their transgressions. He looks on all men who do not conform to his ideas of morality and religion as being sinners and transgressors and thanks God for his own insight and keenness. But he is not at peace. Something seems lacking. God does not talk to him "face to face," because the mind, where God and man meet, is darkened by the murky thought that other men are sinners. Our first work in any demonstration is to contact God, therefore we must forgive all men their transgressions. Through this forgiveness we cleanse our mind so that the Father can forgive us our own transgressions.

Our forgiving "all men" includes ourselves. You must also forgive yourself. Let the finger of denial erase every sin or "falling short" that you have charged up against yourself. Pay your debt by saying to that part of yourself which you think has fallen short: "Thou art made whole: sin no more, lest a worse thing befall thee." Then "loose him, and let him go." Treat sin as a mental transgression, instead of considering it as a moral deflection. Deny in thought all tendency to the error way and hold yourself firmly to the Christ Spirit, which is your divine self. Part company forever with "accusing conscience." Those who have resolved to sin no more have nothing in common with guilt.

"Shall I be in debt as long as I hold debts against others?" We find this to be the law of mind: a thought of debt will produce debt. So long as you believe in debt you will go into debt and accumulate the burdens that follow that thought. Whoever has not forgiven all men their debts is likely to fall into debt himself. Does this mean that you should give receipted bills to all those who owe you? No. That would not be erasing the thought of debt from your mind. First deny in mind that any man or woman owes you anything. If necessary, go over your list of names separately and sincerely forgive the thought of debt which you have been attaching to each person named. More bills may be collected in this way than in any other, for many of these people will pay what they owe when you send them this forgiving thought.

Debt is a contradiction of the universal equilibrium, and there is no such thing as lack of equilibrium in all the universe. Therefore in Spirit and in Truth there is no debt. However, men hold on to a thought of debt, and this thought is responsible for a great deal of sorrow and hardship. The true disciple realizes his supply in the consciousness of omnipresent, universally possessed abundance. Spirit substance is impartial and owned in common, and no thought of debt can enter into it.

Debts exist in the mind, and in the mind is the proper place to begin liquidating them. These thought entities must be abolished in mind before their outer manifestations will pass away and stay away. The world can never be free from the bondage of financial obligations until men erase from their minds the thoughts of "mine and thine" that generates debts and interest. Analyze the thought of debt and you will see that it involves a thought of lack. Debt is a thought of lack with absence at both ends; the creditor thinks he lacks what is owed him and the debtor thinks he lacks what is necessary to pay it, else he would discharge the obligation rather than continue it. There is error at both ends of the proposition and nothing in the middle. This being true, it should be easy to dissolve the whole thought that anyone owes us or that we owe anyone anything. We should fill our mind with thoughts of all-sufficiency, and where there is no lack there can be no debts. Thus we find that the way to pay our debts is by filling our mind with the substance of ideas that are the direct opposite of the thoughts of lack that caused the debts.

Ideas of abundance will more quickly and surely bring what is yours to you than any thoughts you can hold about debtors discharging their obligations to you. See substance everywhere and affirm it, not only for yourself but for everyone else. Especially affirm abundance for those whom you have held in the thought of owing you. Thus you will help them pay their debts more easily than if you merely erased their names from your book of accounts receivable. Help pay the other fellow's debts by forgiving him his debts and declaring for him the abundance that is his already in Spirit. The idea of abundance will also bring its fruits into your own life. Let the law of plenty work itself out in you and in your affairs. This is the way the Father forgives your debts: not by canceling them on His books but by erasing them from His mind. He remembers them no more against you when you deny their reality. The Father is the everywhere present Spirit in which all that appears has its origin. God's love sees you always well, happy, and abundantly provided for; but God's wisdom demands that order and right relation exist in your mind before it may become manifest in your affairs as abundance. His love would give you your every desire, but His wisdom ordains that you forgive your debtors before your debts are forgiven.

To remedy any state of limited finances or ill-health that has been brought about by worry one must begin by eliminating the worry that is the original cause. One must free one's mind from the burden of debt before the debt can be paid. Many people have found that the statement "I owe no man anything but love" has helped them greatly to counteract this thought of debt. As they used the words their minds were opened to an inflow of divine love and they faithfully co-operated with the divine law of forgiveness in thought, word, and deed. They built up such a strong consciousness of the healing and enriching power of God's love that they could live and work peacefully and profitably with their associates. Thus renewed constantly in health, in faith, and in integrity, they were able to meet every obligation that came to them.

The statement "I owe no man anything but love" does not mean that we can disclaim owing our creditors money or try to evade the payment of obligations we have incurred. The thing denied is the burdensome thought of debt or of lack. The work of paying debts is an inner work having nothing to do with the debts already owed but with the wrong thoughts that produced them. When one holds to the right ideas, burdensome debts will not be contracted. Debts are produced by thoughts of lack, impatient desire, and covetousness. When these thoughts are overcome, debts are overcome, forgiven, and paid in full, and we are free from them for all time.

Your thoughts should at all times be worthy of your highest self, your fellow man, and God. The thoughts that most frequently work ill to you and your associates are thoughts of criticism and condemnation. Free your mind of them by holding the thought "There is now no condemnation in Christ Jesus." Fill your mind with thoughts of divine love, justice, peace, and forgiveness. This will pay your debts of love, which are the only debts you really owe. Then see how quickly and easily and naturally all your outer debts will be paid and all lack of harmonies of mind, body, and affairs smoothed out at the same time. Nothing will so quickly enrich your mind and free it from every thought of lack as the realization of divine love. Divine love will quickly and perfectly free you from the burden of debt and heal you of your physical infirmities, often caused by depression, worry, and financial fear. Love will bring your own to you, adjust all misunderstandings, and make your life and affairs healthy, happy, harmonious, and free, as they should be. Love indeed is the "fulfillment of the law."

The way is now open for you to pay your debts. Surrender them to God along with all your doubts and fears. Follow the light that is flooding into your mind. God's power, love, and wisdom, are here, for His kingdom is within you. Give Him full dominion in your life and affairs. Give Him your business, your family affairs, your finances, and let Him pay your debts. He is even now doing it, for it is His righteous desire to free you from every burden, and He is leading you out of the burden of debt, whether of owing or being owed. Meet every insidious thought, such as "I can't," "I don't know how," "I can't see the way," with the declaration "Jehovah is my shepherd; I shall not want." You "shall not want" the wisdom, the courage to do, or the substance to do with when you have once fully realized the scope of the vast truth that Almightiness is leading you into "green pastures ... beside still waters."

In the kingdom of Truth and reality ideas are the coin of the realm. You can use the new ideas that divine wisdom is now quickening in your mind and start this very moment to pay your debts. Begin by thanking God for your freedom from the debt-burden thought. This is an important step in breaking the shackles of debt. The funds to pay all your bills may not suddenly appear in a lump sum; but as you watch and work and pray, holding yourself in the consciousness of God's leadership and His abundance, you will notice your funds beginning to grow "here a little, there a little," and increasing more and more rapidly as your faith increases and your anxious thoughts are stilled. For with the increase will come added good judgment and wisdom in the management of your affairs. Debt is soon vanquished when wisdom and good judgment are in control.

Do not yield to the temptation of "easy-payment plans." Any payment that drains your pay envelope before you receive it is not an easy payment. Do not allow false pride to tempt you to put on a thousand-dollar front on a hundred-dollar salary. There may be times when you are tempted to miss paying a bill in order to indulge a desire for something. This easily leads one into the habit of putting off paying, which fastens the incubus of debt on people before they realize it. It is the innocent-appearing forerunner of the debt habit and debt thought that may rob you of peace, contentment, freedom, integrity, and prosperity for years to come. The Divine Mind within you is much stronger than this desire mind of the body. Turn to it in a time like this, and affirm: "Jehovah is my shepherd; I shall not want" this thing until it comes to me in divine order.

Bless your creditors with the thought of abundance as you begin to accumulate the wherewithal to pay off your obligations. Keep the faith they had in you by including them in your prayer for increase. Begin to free yourself at once by doing all that is possible with the means you have and as you proceed in this spirit the way will open for you to do more; for through the avenues of Spirit more means will come to you and every obligation will be met.

If you are a creditor, be careful of the kind of thoughts you hold over your debtor. Avoid the thought that he is unwilling to pay you or that he is unable to pay you. One thought holds him in dishonesty, and the other holds him subject to lack, and either of them tends to close the door to the possibility of his paying you soon. Think well and speak well of all those who owe you. If you talk about them to others avoid calling them names that you would not apply to yourself. Cultivate a genuine feeling of love for them and respect their integrity in spite of all appearances. Declare abundant supply for them and thus help them to prosper. Pray and work for their good as well as for your own, for yours is inseparable from theirs. You owe your debtor quite as much as he owes you and yours is a debt of love. Pay your debt to him and he will pay his to you. This rule of action never fails.

Far-seeing Christians look forward to an early resumption of the economic system inaugurated by the early followers of Jesus. They had all things in common, and no man lacked anything. But before we can have a truly Christian community founded on a spiritual basis we must be educated into a right way of thinking about finances. If we should all get together and divide all our possessions, it would be but a short time until those who have the prevailing financial ideas would manipulate our finances, and plethora on one hand and lack on the other would again be established.

The world cannot be free from the bondage of debt and interest until men start to work in their minds to erase those things from consciousness. If the United States forgave the nations of Europe all their debts and wiped the slate clean, the law would not necessarily be fulfilled; for there would probably remain a thought that they still owed us and that we had made a sacrifice in canceling the obligations. We should not feel very friendly about it and would not truly forgive them, and in that case the error thought would be carried on. We must first forgive the error thought that they owe us money and that we would be losing money by canceling the debts. The man who is forced to forgive a debt does not forgive it.

Above all we should fill our mind with the consciousness of that divine abundance which is so manifest everywhere in the world today. There is as much substance as there ever was, but its free flow has been interfered with through selfishness. We must rid our mind of the selfish acquisitiveness that is so dominant in the race thought, and in that way do

our part in the great work of freeing the world from avarice. It is the duty of every Christian metaphysician to help in the solution of this problem by affirming that the universal Spirit of supply is now becoming manifest as a distributing energy the world over; that all stored-up, hoarded, vicious thoughts are being dissolved; that all people have things in common. that no one anywhere lacks anything; and that the divine law of distribution of infinite supply that Jesus demonstrated is now being made manifest throughout the world. "The earth is the Lord's, and the fullness thereof."

There is a legitimate commerce that is carried on by means of what is called credit. Credit is a convenience to be used by those who appreciate its value and are careful not to abuse it, for to do so would be to ruin it. However, many persons are not equipped to use the credit system to advantage and are likely to abuse it. In the first place, few individuals are familiar with the intricacies of sound credit systems and often assume obligations without being certain of their ability to meet them, especially should some unforeseen complication arise. Frequently an individual loses all that he invests and finds himself involved in a burden of debt in addition. Such things are not in divine order and are largely responsible for retarding prosperity.

No one should assume an obligation unless he is prepared to meet it promptly and willingly when it comes due. One who knows God as his unfailing resource can be assured of his supply when it is needed.

Then why should he plunge into debt when he is confident of his daily supply without debt? There are no creditors or debtors in God's kingdom. If you are in that kingdom, you need no longer be burdened with the thought of debt either as debtor or creditor. Under divine law there is no reaching out for things that are beyond one's present means. There is an ever-increasing richness of consciousness coming from the certain knowledge that God is infinite and unfailing supply. Outer things conform to the inner pattern, and riches are attracted to the one who lives close to the unselfish heart of God. His environment is made beautiful by the glory of the Presence, and there is satisfying and lasting prosperity in his affairs.

There is but one way to be free from debt. That is the desire to be free, followed by the realization that debt has no legitimate place in God's kingdom and that you are determined to erase it entirely from your mind. As you work toward your freedom you will find it helpful to have daily periods for meditation and prayer. Do not concentrate on debts or spoil your prayers by constantly thinking of debts. Think of that which you want to demonstrate, not that from which you seek freedom. When you pray, thank the Father for His care and guidance, for His provision and plenty, for His love and wisdom, for His infinite abundance and your privilege to enjoy it.

Here are a few prosperity prayers that may help establish you in the truth of plenty and erase the error thought of debt. They are offered as suggestions for forming your own prayers but may be used as given with excellent results.

I am no longer anxious about finances; Thou art my all-sufficiency in all things.

The Spirit of honesty, promptness, efficiency, and order is now expressed in me and in all that I do.

I am free from all limitations of mortal thought about quantities and values. The superabundance of riches of the Christ Mind are now mine, and I am prospered in all my ways.

The 23d Psalm

A Treatment To Free The Mind Of The Debt Idea

Jehovah is my shepherd; I shall not want

He maketh me to lie down in green pastures:

He leadeth me beside still waters.

He restoreth my soul:

He guideth me in the paths of righteousness for his name's sake.

Yea, though I walk through the valley of the shadow of death,

I will fear no evil; for thou art with me;

Thy rod and thy staff, they comfort me.

Thou preparest a table before me in the presence of mine enemies:

Thou hast anointed my head with oil;

My cup runneth over.

Surely goodness and lovingkindness shall follow me all the days of my life:

And I shall dwell in the house of Jehovah for ever.

Lesson 9
Tithing, the Road to Prosperity

As ye abound in everything, in faith, and utterance, and knowledge, and in all earnestness, and in your love to us, see that ye abound in this grace also."

"Honor Jehovah with thy substance,

And with the first-fruits of all thine increase:

So shall thy barns be filled with plenty,

And thy vats shall overflow with new wine."

Under the Mosaic law a tithe (or tenth) was required as the Lord's portion. Throughout the Old Testament the tithe or tenth is mentioned as a reasonable and just return to the Lord by way of acknowledging Him as the source of supply. After Jacob had seen the vision of the ladder with angels ascending and descending on it he set up a pillar and made a vow to the Lord, saying, "Of all that thou shalt give me I will surely give the tenth unto thee." In the 3d chapter of Malachi we find God's blessing directly connected with faithfulness in giving to the Lord's treasury, but gifts should be made because it is right and because one loves to give, not from a sense of duty or for the sake of reward.

That there will be a reward following the giving we are also assured by Jesus in a direct promise "Give, and it shall be given unto you; good measure, pressed down, shaken together, running over, shall they give unto your bosom. For with what measure ye mete it shall be measured to you again."

Promises of spiritual benefits and increase of God's bounty through the keeping of this divine law of giving and receiving, abound in all the Scriptures. "There is that scattereth, and increaseth yet more;

And there is that withholdeth more than is meet, but it tendeth only to want.

The liberal soul shall be made fat;

And he that watereth shall be watered also himself."

"He that hath a bountiful eye shall be blessed;

For he giveth of his bread to the poor."

"He that soweth bountifully shall reap also bountifully."

"Blessed are ye that sow beside all waters."

We are living now under larger and fuller blessings from God than man has ever known. It is meet therefore that we give accordingly and remember the law of the tithe, for if a tenth was required under the law in those olden times, it is certainly no less fitting that we should give it cheerfully now. One of the greatest incentives to generous giving is a keen appreciation of the blessings secured to us through the redemptive work of Jesus Christ. "He that spared not His own Son, but delivered him up for us all, how shall he not also with him freely give us all things?" "Freely ye received, freely give." True giving is the love and generosity of the Spirit-quickened heart responding to the love and generosity of the Father's heart.

In his second letter Paul made a stirring appeal to the Corinthians for a generous gift to their poorer brethren in Jerusalem. He suggests some principles of giving that are always applicable, for giving is a grace that adds to the spiritual growth of all men in all times. Without giving the soul shrivels, but when giving is practiced as a part of Christian living, the soul expands and becomes Godlike in the grace of liberality and generosity. No restoration to the likeness of God can be complete unless mind, heart, and soul are daily opening out into that large, free, bestowing spirit which so characterizes our God and Father. Therefore it is not surprising that Paul classes the grace of giving with faith, knowledge, and love.

A very simple yet practical plan for exercising this grace of giving had been suggested by Paul in his first letter to the Corinthian church. "Upon the first day of the week," he said, "let each one of you lay by him in store, as he may prosper"; that is, each member was asked to contribute to the establishing of a treasury. This was to be the Lord's storehouse, into which each one was to put his offerings regularly and in proportion to his means. In adopting this plan the offerer became a steward of the Lord's goods and entered upon a course of training and discipline needed to make a good steward, for it takes wisdom to know how rightly to dispense the bounty of God. Perhaps no simpler way to begin one's growth in the grace of giving can be suggested for our own day. Those who have followed this method have usually found that they had more money to give than they had thought possible.

In order that the plan of giving may be successful there are several things that must be observed. First there must be a willing mind. "If the readiness is there, it is acceptable according as a man hath, not according as he hath not." "God loveth a cheerful giver." Secondly, the giving must be done in faith, and there must be no withholding because the offering seems small. Many of the instances of giving that are recorded in the Bible as worthy of special mention, commendation, and blessing are instances where the gift itself was small. The widow who fed Elijah in his time of famine gave him a cake made with her last handful of meal. For her faith and her generous spirit she was rewarded with a plentiful daily supply of food for herself and her sons, as well as for Elijah. "The jar of meal shall not waste, neither shall the cruse of oil fail."

This same truth is set forth beautifully in the New Testament, where it is clearly shown that not the amount of the offering but the spirit in which it is given determines its value and power. "And he [Jesus] sat down over against the treasury, and beheld how the multitude cast money into the treasury: and many that were rich cast in much. And there came a poor widow, and she cast in two mites, which make a farthing. And he called unto him his disciples, and said unto them, Verily I say unto you, This poor widow cast in more than all they that are casting into the treasury: for they all did cast in of their superfluity; but she of her want did cast in all that she had, even all her living."

This poor widow exemplified what it is to give in faith; and were ever two mites so great a gift as when they brought forth such praise from the Master Himself! The results of giving in faith are just as sure in this age as in the time of Jesus, for the law is unfailing in all ages.

A third requisite for keeping the law of giving and receiving is that the offering shall be a just and fair proportion of all that one receives. The amount was settled by Paul and the measure he gave was: "as he may prosper." There is a certain definiteness about this, and yet it admits of freedom for the giver to exercise his individual faith, judgment, and will.

The question of wise distribution is closely related to the matter of filling God's treasury. To whom shall we give and when are questions quite important. There are several truths that may be considered in this connection, but then each individual finds it necessary to trust to the Spirit of wisdom manifest in his own heart, since there are no rules or precedents that one can follow in detail. This is as it should be, for it keeps the individual judgment, faith, love, sympathy, and will alive and active. Yet a careful study of the underlying laws of spiritual giving will help one to exercise these divine faculties as they should be exercised. If we follow the Spirit of wisdom we shall not give to anything that is contrary to the teaching of Jesus, but spend every penny in the furtherance of the good news of life that He proclaims and in the promotion of the brotherhood of man that it is His mission to establish on earth among all those who become sons through Him.

True spiritual giving rewards with a double joy: first that which comes with the laying of the gift upon the altar or in the Lord's treasury; then the joy of sharing our part of God's bounty with others. One of the blessings is the satisfying knowledge that we are meeting the law and paying our debt of love and justice to the Lord. The other is the joy of sharing the Lord's bounty. Justice comes first; then generosity.

Even the so-called heathen recognize giving as a part of worship, for we find them coming with offerings when they worship their idols. All ages and all religious dispensations have stressed giving as a vital part of their worship. In this age, when we have so much, more is required of us, even to the giving of ourselves with all that we are and have. This privilege carries immeasurable benefits with it, for it looses us from the personal life, unifies us with the universal, and so opens our inner and outer life to the inflow and the outflow of the life, love, bounty and grace of God. This is the blessed result of faithful obedience to the law and exercise of the grace of giving.

The people were amazed when the prophet Malachi told them that they had been robbing God and desired to know wherein they had failed when they thought they had been serving the Lord so faithfully. People are as much amazed today to learn that they have been untrue to God's law, for the message of Malachi is for us quite as much as for the ancients. The Spirit of God gave this message through the prophet: "Bring ye the whole tithe into the store-house, that there may be food in my house, and prove me now herewith, saith Jehovah of hosts, if I will not open you the windows of heaven, and pour you out a blessing, that there shall not be room enough to receive it. And I will rebuke the devourer for your sakes, and he shall not destroy the fruits of your ground; neither shall your vine cast its fruit before the time in the field, saith Jehovah of hosts. And all nations shall call you happy; for ye shall be a delightsome land, saith Jehovah of hosts."

Study this 3d chapter of Malachi carefully if you would know the happy solution of the problem of giving and receiving. See how practical it is for people in every walk of life and for nations as well. It offers the solution to the problems of the farmer. It sets forth clearly a law of prosperity for all classes of people; for those who need protection for their crops from frosts, droughts, floods; for those who would escape the plagues, pestilences, and manifold things that would destroy their supply and support. It is a simple law but so effective: simply give a tithe or tenth or the "first-fruits" or their equivalent to the Lord. God should not be expected to meet all man's requirements in the matter of giving this protection and increase unless man fulfills the requirements of God. The act of giving complies with the divine law, because it involves the recognition of God as the giver of all increase; and unless we have a recognition of the source of our supply we have no assurance of continuing in its use.

Many people have doubts as to whether it will really do any good to ask the Lord for protection and for plenty in regard to crops or other supply. Many who are employed in cities or who are in business think it strange that they should believe in omnipresent prosperity. Thus unbelief is present with them at the very time when an unwavering faith is most necessary. There is a psychological reason why people should obey spiritual law. When a person obeys the law of God along any line, his faith immediately becomes strengthened in proportion and his doubts disappear. When anyone puts God first in his finances, not only in thought but in every act, by releasing his first fruits (a tenth part of his increase or income) to the Lord, his faith in omnipresent supply becomes a hundredfold stronger and he prospers accordingly. Obeying this law gives him an inner knowing that he is building his finances on a sure foundation that will not fail him.

Everything in the universe belongs to God, and though all things are for the use and enjoyment of man, he can possess nothing selfishly. When man learns that a higher law than human custom and desire is working in the earth to bring about justice, righteousness, and equalization, he will begin to obey that law by tithing, loving his neighbor, and doing unto others as he would have them do unto him. Then man will reach the end of all the troubles brought upon himself by his selfishness and greed, and will become healthy, prosperous, and happy.

The pastor of a small church in Georgia suggested to his congregation, composed largely of cotton farmers, that they dedicate a tenth part of their land to the Lord and ask Him for protection against the ravages of the boll weevil, which had devastated the crops in that vicinity for several years. Seven farmers in the congregation decided to do this. They took no measures to protect their crop on these dedicated acres, yet the pest did not attack the cotton there. The quality of the fiber was better on those acres than on any that adjoined them. The experiment was so successful in fact that practically all the farmers in that community have decided to follow the plan in the future.

Many experiences such as this are awakening men to respect our relation to the infinite principle of life, everywhere present, that we know as God. This divine element of life that manifests itself as growth and substance resides within the factors that combine to produce cotton, wheat, and all other forms of vegetation. Then certainly if the farmer works in acknowledged sympathy with this life principle, it will work in sympathy with him and for his good. Each contributing in love and understanding to the other, a larger crop will be the result, and a larger measure of prosperity for the farmer. Not only the farmer but the banker, the tradesman, the professional man can work in sympathy and harmony with this principle of growth and increase. The infinite life principle is as responsive in one field as another, and it is everywhere present. Even so-called inanimate objects are filled to the full with this infinite life, and even coined gold is tense with the desire to expand and to grow. The materials handled by the tradesman are made of the same substance that makes the universe and contain within themselves the germ of growth and increase. All men are therefore daily associated with life, and through rendering it the reverent acknowledgment that is its due and through witnessing this acknowledgment by dedicating a part of their increase they are prospered.

The tithe is the equivalent of the increased fertility of the land. If by acknowledging God as the giver of all life the farmer raises two or six or twenty bushels more on his field, that extra portion, which he would not have had otherwise, is the Lord's portion. In trade the tithe is the equivalent of the increased quality of goods. In professional life the tithe is the increased ability or the increased appreciation. The tithing principle can be applied in all of our industrial and social relationships. In every case where it has been applied and followed for a time, the tither has been. remarkably blessed; quite as much so as in the case of the cotton farmers and their tithe acres.

There are many people who wish to give but seem at a loss as to how to go about it or where to begin. They do not know how much they should give, or when or how often to offer their gifts, and there are a host of related questions. To answer these questions there must be found a definite basis for their giving, a rule to which they can conform. This is where the law of tithing fits beautifully, for it is a basis and a sound one, tested and proved for thousands of years. The tithe may be a tenth part of one's salary, wage, or allowance, of the net profits of business, or of money received from the sale of goods. It is based on every form of supply, no matter through what channel it may come, for there are many channels through which man is prospered. The tenth should be set apart for the upkeep of some spiritual work or workers. It should be set apart first even before one's personal expenses are taken out, for in the right relation of things God comes first always. Then everything else follows in divine order and falls into its proper place.

The great promise of prosperity is that if men seek God and His righteousness first, then all shall be added unto them. One of the most practical and sensible ways of seeking God's kingdom first is to be a tither, to put God first in finances. It is the promise of God, the logical thing to do, and the experience of all who have tried it, that all things necessary to their comfort, welfare, and happiness have been added to them in an overflowing measure. Tithing establishes method in giving and brings into the consciousness a sense of order and fitness that will be manifested in one's outer life and affairs as increased efficiency and greater prosperity.

Another blessing that follows the practice of tithing is the continual "letting go" of what one receives, which keeps one's mind open to the good and free from covetousness. Making an occasional large gift and then permitting a lapse of time before another is made will not give this lasting benefit, for one's mind channel may in the meantime become clogged with material thoughts of fear, lack, or selfishness. When a person tithes he is giving continuously, so that no spirit of grasping, no fear, and no thought of limitations gets a hold on him. There is nothing that keeps a person's mind so fearless and so free to receive the good constantly coming to him as the practice of tithing. Each day, week, pay day, whenever it is, the tither gives one tenth. When an increase of prosperity comes to him, as come it will and does, his first thought is to give God the thanks and the tenth of the new amount. The free, open mind thus stayed on God is certain to bring forth joy, real satisfaction in living, and true prosperity. Tithing is based on a law that cannot fail, and it is the surest way ever found to demonstrate plenty, for it is God's own law and way of giving.

"And all the tithe of the land, whether of the seed of the land, or of the fruit of the tree, is Jehovah's: it is holy unto Jehovah."

Let us give as God gives, unreservedly, and with no thought of return, making no mental demands for recompense on those who have received from us. A gift with reservations is not a gift; it is a bribe. There is no promise of increase unless we give freely, let go of the gift entirely, and recognize the universal scope of the law. Then the gift has a chance to go out and to come back multiplied. There is no telling how far the blessing may travel before it comes back, but it is a beautiful and encouraging fact that the longer it is in returning, the more hands it is passing through and the more hearts it is blessing. All these hands and hearts add something to it in substance, and it is increased all the more when it does return.

We must not try to fix the avenues through which our good is to come. There is no reason for thinking that what you give will come back through the one to whom you gave it. All men are one in Christ and form a universal brotherhood. We must put away any personal claim, such as "I gave to you, now you give to me," and supplant it with "Inasmuch as ye did it unto one of these my brethren, even these least, ye did it unto me." The law will bring each of us just what is his own, the reaping of the seeds he has sown. The return will come, for it cannot escape the law, though it may quite possibly come through a very different channel from what we expect. Trying to fix the channel through which his good must come to him is one of the ways in which the personal man shuts off his own supply.

The spiritual-minded man does not make selfish use of the law but gives because he loves to give.

Because he gives with no thought of reward and no other motive than love, he is thrown more completely into the inevitable operation of the law and his return is all the more certain. He is inevitably enriched and cannot escape it. Jesus said, "Give, and it shall be given unto you; good measure, pressed down, shaken together, running over." He was not merely making a promise but stating a law that never fails to function.

So inexhaustible is the bounty of the Giver of all good that to him who has eyes to see it and faith to receive it God is an unfailing source of supply. The munificent Giver withholds nothing from him who comes in the name of a son and heir and lays claim to his portion. It is the Father's good pleasure to give us the kingdom, and all that the Father has is ours. But we must have the faith and the courage to claim it.

Men who accomplish great things in the industrial world are the ones who have faith in the money-producing power of their ideas. Those who would accomplish great things in the demonstration of spiritual resources must have faith to lay hold of the divine ideas and the courage to speak them into expression. The conception must be followed by the affirmation that the law is instantly fulfilled. Then the supply will follow in manifestation.

Lesson 10
Right Giving, the Key to Abundant Receiving

There is a law of giving and receiving and it requires careful study if we would use it in our prosperity demonstrations. It is a law of mind action, and it can be learned and applied the same as any other law. The teaching of Jesus stands out prominently, because it can be practically applied to the affairs of everyday life. It is not alone a religion in the sense that word is usually taken but is a rule of thinking, doing, living, and being. It is not only ethical but practical, and men have never yet sounded the depths of the simple but all-inclusive words of Jesus. To some people it is unthinkable to connect the teaching of Jesus with the countinghouse and the market place, but a deeper insight into their meaning and purpose, which the Spirit of Truth is now revealing to the world, shows that these lofty teachings are the most practical rules for daily living in all departments of life. They are vital to modern civilization and the very foundation of business stability. The law of giving and receiving that Jesus taught, "Give, and it shall be given unto you," is found to be applicable to all our commercial as well as our social relationships.

We have not been more successful in making this doctrine of Jesus a practical standard for everyday guidance because we have not understood the law on which it is based. Jesus would not have put forth a doctrine that was not true and not based on unchanging law, and we can be sure that this doctrine of giving and receiving is powerful enough to support all the affairs of civilization. We have not gone deeply enough into the teaching but have thought we understood it from a mere surface study. "Ye look at the things that are before your face," says Paul, and Jesus also warned us to "judge not according to appearance." We should form no conclusions until we have gone thoroughly into the causes and the underlying laws. The things we see outwardly are the effects that have arisen from causes that are invisible to us. There is an inner and an outer to everything: both the mental and the material conditions pervade the universe. Man slides at will up and down the whole gamut of cause and effect. The whole race slides into an effect almost unconsciously and so identifies the senses with the effect that the causes are lost sight of for thousands of years.

An awakening comes in time and the cause side of existence is again brought to the attention of men, as set forth, for example, in the doctrine of Jesus Christ. But men cannot grasp the great truth in a moment and cling to what is plainly visible to them, the effect side. The truth that things have a spiritual as well as a material identity and that the spiritual is the cause side and of greatest value, is a revelation that may be slow in coming to most people. In this instance it is the material side that they cling to, thinking it to be all and refusing to let go. Men have taken the letter or appearance side of the Jesus Christ doctrine and materialized it to fit their beliefs and customs. That is the reason why the Christ message has not purified commerce, society, and government. But it should be made spiritually operative in those fields. It will easily do the work desired when its mental side is studied and when it is understood and applied from the spiritual viewpoint.

There is need for reform in economics more than in any other department of everyday life. Money has been manipulated by greed until greed itself is sick and secretly asks for a panacea. But it does not look to the religion of Jesus Christ for healing. In fact that is the very last place it would apply for aid, because many of the advocates of the Jesus Christ doctrine are themselves economic dependents and have no solution for the economic problem--not understanding the power of their own religion. Yet no permanent remedy will ever be found for the economic ills of the world outside a practical application of the laws on which the doctrine of Jesus Christ is based.

The correctness of the solution of any problem is assured by the right relation of its elements. All true reform begins with the individual. Jesus began there. He did not clamor for legislation to control men or their actions. He called His twelve apostles and through them individually instituted that reform which has as its basis an appeal to the innate intelligence, honesty, and goodness in every man. He told them, "Go ye into all the world, and preach the gospel to the whole creation."

As people learn more definitely about the dynamic effect of thought and how ideas pass from mind to mind, they see more and more the wisdom of the Christ teaching. They are beginning to understand that there is one undeviating law of mind action and that all thinking and all speaking is amenable to it. Thus when Jesus said, "By thy words thou shalt be justified, and by thy words thou shalt be condemned," He taught the power of thoughts and words to bring results in accordance with the ideas back of them.

Following the metaphysical side of the teaching of Jesus, we have found that certain thoughts held in the minds of the people are causing widespread misery, disease, and death. We have also found that these thoughts can be dissolved or transformed and the whole man made over through his conscious volition. Paul well understood this process. He said, "Be ye transformed by the renewing of your mind."

Among the destructive thoughts that men indulge in and exercise are those forms of selfishness which we know as avarice, covetousness, money getting, the desire for financial gain and for possessing the things of the world. These thoughts threaten seriously to disturb the civilization of the world and the stability of the whole race. The sole thought of money getting is being allowed by men and women to generate its cold vapor in their souls until it shuts out all the sunlight of love and even of life. The remedy for the misery caused by destructive thoughts is not far to seek. It lies in constructive thinking along the lines that Jesus laid down. Indeed the remedy for all the ills to which flesh is heir lies in conformity to the divine law that Jesus revealed to His true followers. It is said of these true followers (Acts 4:32) that they were "of one heart and soul: and not one of them said that aught of the things which he possessed was his own; but they had all things common."

Many true Christians have observed this righteous law and sought to conform to it in community life. Such efforts have not always been successful, because there was not the necessary recognition of the mental factor and the discipline of ideas. So long as the idea of covetousness is lodged in the human mind as its dominant generating factor, there can be no successful community life. That idea must be eliminated from the mental plane first; the next step, the outer practice, will then be safe and successful.

Everywhere true metaphysicians are preparing themselves to be members in the great colony that Jesus is to set up, by working to eliminate from their mind all selfish ideas, along with all other discordant vibrations that produce disharmony among members of the same group. A step in this direction is the gradual introduction of the "freewill offering" plan to replace the world's commercial standard of reward for services. We are striving to educate the people on this question of giving and receiving and to let their own experience prove to them that there is a divine law of equilibrium in financial matters that corresponds to the law of balance and poise that holds the suns and the planets in place. In order to make a success of this great effort we must have the loving co-operation of everyone to whom we minister. The law is based on love and justice, and it equitably and harmoniously adjusts all the affairs of men. It goes even further, for it restores a harmony and balance in both mind and body that results in happiness and health as well as prosperity. Love and justice are mighty powers, and all things must eventually come under their influence, because even a few men and women of right motive can, by right thinking and consequent just action, introduce these ideas into the race consciousness and pave the way for their universal adoption. The movement has already begun and is rapidly gaining headway. Every student and reader is asked to give it impetus by resolving to be unselfish and just without compulsion.

The race consciousness is formed of thought currents and the dominant beliefs of all the people. A few men and women rise above these currents of thought and become independent thinkers. The dominating race idea of money getting as the goal of success is now being replaced by the idea of usefulness and good works. This idea must be carried

out by individuals who have resolved to think and to act in the Jesus Christ way. To be one of these individuals and to contribute to the change in the race consciousness, first dedicate yourself in Spirit to the ministry of Jesus and resolve to carry forward the great work He has commissioned you to do. This does not mean that you must preach like Paul or necessarily carry on any extensive work in the outer. In the silence of your "inner chamber" you can do a mighty work of power by daily denying the beliefs in avarice and covetousness and affirming the universal sway of divine love and justice. You can make the idea of exact equity and justice between man and man the central theme of all your saying and doing. When you see examples of greed and avarice or when thoughts of these seek a place in your mind, remember the words of the Master: "What is that to thee? follow thou me."

Never for a moment allow yourself to entertain any scheme for getting the better of your fellows in any trade or bargain. Hold steadily to the law of equity and justice that is working in and through you, knowing for a certainty that you are supplied with everything necessary to fulfill all your requirements. Give full value for everything you get. Demand the same for everything you give, but do not try to enforce that demand by human methods. There is a better way: think of yourself as Spirit working with powerful spiritual forces, and know that the demands of Spirit must and will be met.

Do not plan to lay up for the future; let the future take care of itself. To entertain any fears or doubts on that point saps your strength and depletes your spiritual power. Hold steadily to the thought of the omnipresence of universal supply, its perfect equilibrium and its swift action in filling every apparent vacuum or place of lack. If you have been in the habit of hoarding or of practicing stringent economy, change your thought currents to generosity. Practice giving, even though it may be in a small way. Give in a spirit of love and give when you cannot see any possibility of return. Put real substance into your gift by giving the substance of the heart with the token of money or whatever it is. Through the power of your word you can bless and spiritually multiply everything that you give. See yourself as the steward of God handing out His inexhaustible supplies. In this manner you are setting into action mental and spiritual forces that eventually bring large results into visibility. Be happy in your giving. God loves a cheerful giver because his mind and heart are open to the flow of the pure substance of Being that balances all things.

Do not give with any idea that you are bestowing charity. The idea of charity has infested the race consciousness for thousands of years and is responsible for the great army of human dependents. Do all you can to annul this mental error. There is no such thing as charity as popularly understood. Everything belongs to God and all His children are equally entitled to it. The fact that one has a surplus and gives some of it to another does not make the one a benefactor and the other a dependent. The one with the surplus is simply a steward of God and is merely discharging the work of his stewardship. When one asks for divine wisdom and understanding about giving it becomes a joy both to the giver and the recipient.

Followers of Jesus who are doing His work of teaching and healing should, like Him, receive free-will offerings for their ministry to the people. The majority of those who apply to teachers and healers recognize this law of giving and receiving, but there are quite a number who do not understand it. First there are those who are in bondage to the idea of avarice, and secondly, there are those who still are in bondage to the idea of charity. Both these classes need education and treatment to release them from mental limitation and mental disease. The avaricious suffer most in body and are the most difficult to heal, because of the mental bias that prompts them to get everything as cheaply as possible, including the kingdom of heaven. They must be patiently educated to be just because it is right, and to learn to "let go" of the acquisitive spirit and replace it with the spirit of generosity. They will do this readily enough as a mental drill but are not so willing to let go of the money symbol. However, continued treatments in the silence, supplemented with oral and written instruction, will eventually prevail and heal them.

There are many examples that could be given to prove the outworking of the law. The covetous idea has a great power over the body. It would avail little to treat the outer manifestation before first removing the inner cause from the mind. The salvation of such people is to learn to give generously and freely, not from compulsion or for the sake of reward but from a love of the giving. Some metaphysicians think to cure their patients of the hold of avaricious thoughts by charging them a good round price for their treatments. By the same token the medical doctor who charges the most is surest to heal his patients, and any service for which an exorbitant price is charged is the best! Surely this would be a foolish idea. Metaphysical healing has become so popular that hundreds have gone into it as a business and are making of it an industry founded on the old commercial idea, just as cold and calculating, as hard and unyielding as the idea is in the ranks of the money-changers of mammon.

Surely there is a "more excellent way," one more in harmony with divine law, a way that permits the heart as well as the head and hand to be used in the grace of giving and receiving. Those who are using the freewill offering method meet with some criticism and opposition from those who hold to the commercial method and say that charging a definite sum is the legitimate way. They accuse Unity of fostering charity and poverty and keeping alive the spirit of getting something for nothing that is manifested by so many people. Our reply is that we are pursuing the only course that could ever effectually eradicate these erroneous states of consciousness and bring people into an understanding of the spiritual law of prosperity through giving in love.

Everyone should give as he receives; in fact, it is only through giving that he can receive. Until the heart is quickened at the center and the mind is opened up to Truth there is no permanent healing. Everyone can make a fair return for everything he gets. We aim to show moneyless paupers that they can give something in return for the good that has been done them. It may be to pass the true word to some other needy one, or merely to lift up their voice in thanksgiving and praise where before they were dumb. We recognize the necessity of some action of the mammon-bound mind. It must be made to let go somewhere before it can receive the light and the power of Spirit.

Our work is to bring men and women to the place of true and lasting dominion where they are superior to both riches and poverty. We can do this by showing them that they are spiritual beings, that they live in a spiritual world here and now, and that through the apprehension of the Truth of their being and their relation to God this dominion is to be realized.

The central and most vital fact that they must come to realize is that an idea has the power of building thought structures, which in turn materialize in the outer environment and affairs and determine every detail of their existence. Every man is a king ruling his own subjects. These subjects are the ideas existing in his mind, the "subjects" of his thought. Each man's ideas are as varied and show as many traits of character as the inhabitants of any empire. But they can all be brought into subjection and made to obey through the I AM power that is the ruler of the kingdom. In your domain of mind there may be colonies of alien ideas--the Philistines, Canaanites, and other foreign tribes, that the Children of Israel found in their Promised Land when they attempted to take possession of it. The story of the Children of Israel and how they gained the possession of that land is a symbolical representation of the experience of everyone who seeks to reclaim his own consciousness in the name of the Lord. The meaning in Hebrew of the name Canaanite is "merchant" or "trader"; in other words, a set of ideas that has to do with the commercial phase of life. Study the Children of Israel (spiritual ideas) in their experiences with these Canaanites and you will get many valuable hints on subduing and handling your own money-getting ideas.

You may allow avariciousness and stinginess to develop in your mind domain until the very blood in your body starts to dry up and your nerves are shaken and palsied with the fear of future poverty. If so, it is time these ideas were driven out and a new set of ideas settled in your domain to become active in building up a new state of consciousness (nation). Begin at once to let go of your all-consuming thoughts of gain. Think about generosity and begin to be generous for your own sake. "It is more blessed to give than to receive" will prove itself to you as the law, for you will be blessed by a new influx of ideas of life, health, and prosperity when you start giving.

Instead of being grasping and avaricious, perhaps you have gone to the other extreme and have cultivated ideas of small things financially. You may have been fostering poverty by holding ideas of pennies instead of dollars or of hundreds instead of thousands. You may be thinking that you cannot give because your income is small or your supply is limited. Your remedy is to cultivate ideas of abundance. Claim God as your inexhaustible resource; that all things are yours. But in order to set in motion the accumulated energy of your thought you must also begin to give. You may be able to give only pennies at first, but give them in the name and the spirit of your opulent God. Send them forth with all the love of your heart and say to them as they go, "Divine love through me blesses and multiplies you."

Your consciousness is like a stream of water. If the stream is in any way dammed up, the water settles in all the low places and becomes stagnant. The quickest way to purify and reclaim the low, "swampy" places in your consciousness is to let in the flood from above by opening the dam. Many people try to demonstrate God as their supply by repeating affirmations of abundance now present, but fail to deny and thus to let go of the old condition and old belief in lack by beginning to give as generously as possible. It is not the amount you give measured by standards of the world, it is the good will you send forth with the gift; which can be measured only by spiritual standards.

"God loveth a cheerful giver." The Greek word here translated cheerful is hilarion, which means really "hilarious, joyful." The gift may be measured in dollars and cents but God looks not on such standards, He looks on and loves the "joyful" giver. We read in Deuteronomy 28:47,

48, "Because thou servedst not Jehovah thy God with joyfulness, and with gladness of heart, by reason of the abundance of all things; therefore shalt thou serve thine enemies ... in hunger, and in thirst, and in nakedness, and in want of all things." This shows that there is a definite relation between the cheerfulness or joyfulness of our giving and our prosperity. Whether we make a large or a small gift, let us make it with largeness of cheer and joy, even of hilarity, remembering that God loveth a "hilarious" giver. "Keep therefore the words of this covenant, and do them, that ye may prosper in all that ye do."

Blessings That May Be Placed On Our Gifts

Divine love, through me, blesses and multiplies this offering.

The Father gives abundantly; I receive thankfully, and give again generously.

This is the bounty of God, and I send it forth with wisdom and joy.

Divine love bountifully supplies and increases this offering.

I give freely and fearlessly, fulfilling the law of giving and receiving.

Lesson 11
Laying Up Treasures

After the multitude had been fed by the increase of the loaves and fishes, Jesus commanded that they gather up the fragments so that nothing might be lost. "And they all ate, and were filled: and they took up that which remained over of the broken pieces, twelve baskets full." Any form of waste is a violation of the divine law of conservation. Everywhere in nature there is evidence of stored-up energy substance, ready for use when needed.

This reserve force is not material but spiritual. It is ready to be called into expression to meet any need. But when it is not put to use or called into expression, there is a manifestation of disharmony or lack either in the body of man or in its outer supply. It is in his wrong conception of this spiritual force that man makes the mistake of falling into the habit of hoarding instead of conserving. He tries to gather things together in the external in a vain effort to avert an imagined shortage in the future and he counts himself rich by the amount of his material possessions.

Spiritually awakened people are coming to know that all riches are spiritual and within the reach of all as divine ideas. They study the law of conservation as it pertains to the spiritual and seek to build up a large reserve consciousness of substance, life, strength, and power, rather than laying up material treasures that "moth and rust consume" and "thieves break through and steal."

Men and women scatter their energies to the four winds in the effort to satisfy the desires of the flesh, and then wonder why they do not demonstrate prosperity. If they only realized the truth that this same thought force can be conserved and controlled to express itself in constructive channels, they would soon be prosperous. Spirit must have substance to work on and there must be substance in the ideas of your mind. If your substance is going here, there, and everywhere, being spent in riotous thinking, how can it accumulate to the point of demonstration? Such a waste of substance is a violation of the law of conservation, a law that all should know. When you overcome your desire for dissipation, not the overt acts only but the inner desire, then you will begin to accumulate substance that must manifest itself as prosperity according to the law.

One of the fundamental principles in the study of Christianity is that God's great objective is the making of a perfect man. Man is the apex of creation, made in God's image and likeness, and endowed with full authority and dominion over his elemental thoughts. We sometimes think that we must succeed in some business or occupation before we can become rich or famous. This is a missing of the mark of "the high calling of God in Christ Jesus," which is to demonstrate the divine idea of a perfect man. The real object of life is not making money or becoming famous but the building of character, the bringing forth of the potentialities that exist in every one of us. A part of the divine plan is substantial provision by the Creator for all the mental and physical needs of His creation. We are not studying prosperity to become rich but to bring out those characteristics that are fundamental to prosperity. We must learn to develop the faculty that will bring prosperity and the character that is not spoiled by prosperity.

Faith is the faculty of mind that deals with the universal-substance idea. Faith is the substance of things hoped for. Everything in God is ideal, without form or shape but with all possibilities. He is omnipresent in our mind and in our body. It is in our body that we bring God into visible manifestation. Faith is the faculty that does this. It lays hold of the substance idea and makes it visible.

The scramble for wealth seems to be the only object of existence for certain minds. Writers of Biblical times were incessantly preaching against the evils of money. Yet Jehovah was always promising riches and honor to all those who kept His commandments. The gold and silver that God promised were spiritual rather than material. God is mind, and mind can give only ideas. These ideas can be translated

into terms of gold or of anything else we desire, according to our thought. The only treasures that are worth saving are those we lay up in the heavens of the mind. The only gold that can be trusted to bring happiness is the gold of Spirit. Jesus says, "I counsel thee to buy of me gold refined by fire, that thou mayest become rich; and white garments, that thou mayest clothe thyself, that the shame of thy nakedness be not made manifest."

Paul tells us that "the love of money is a root of all kinds of evil." That means of course that by loving money man has in some way limited it. He has not loved the true source of money but has loved the thing rather than the Spirit that it expresses. He has broken the law by trying to grasp the thing and failing to acknowledge the idea that lies back of it. We must know this law, observing it in the handling of money, and make love the magnet of supply instead of becoming entangled in that selfishness and greed which is causing so much disharmony and suffering in the world today. We should know that there is a universal money substance and that it belongs to all of us in all its fullness.

In the parable of the sower Jesus uses a most striking phrase. Part of the good seed was choked out by thorns and the thorns represent the "deceitfulness of riches." Money is indeed a cheat. It promises ease and brings cares; it promises pleasures and pays with pain; it promises influence and returns envy and jealousy; it promises happiness and gives sorrow; it promises permanence and then flies away.

Metaphysically, it is better or at least safer to be poor than to be rich. Jesus taught this in the parable of the rich man and Lazarus. The rich man is pictured in torment, crying for the poor man to give him a drink of water. But if the rich are miserable, the poor who greatly desire to be rich are equally so. Poverty and riches are the two poles of a magnet whose pivot is a belief that the possession of matter will bring joy to the possessor. This belief is a delusion, and those who are attracted by this belief and allow their minds to be hypnotized by the desire for material possessions are to be pitied whether their desire is realized or not.

The real possessor of wealth is the one who feels that all things are his to use and to enjoy yet does not burden himself with the personal possession of anything. Diogenes was a most happy man though he lived in a tub. His philosophy has outlived the influence of the rich and powerful people who were his contemporaries. He walked around with a lantern at midday looking for an honest man, so they seem to have been as rare in his day as in ours.

However, the widespread desire for material possessions indicates that there is somewhere some good in it. The natural man is from the soil, formed of the dust of the ground, and loves his native element. The spiritual man is from above, originating in the heavens of the mind. He is given first place and like Jacob supplants the natural man. Men should not condemn the earth because of this, yet they should not love it to the exclusion of the heavens. They should understand that substance is the day from which the Father makes the body of His people. "Your heavenly Father knoweth that ye have need of all these things ... But seek ye first his kingdom, and his righteousness; and all these things shall be added unto you."

The divine law holds that the earth is the Lord's and the fullness thereof. If this truth were thoroughly understood, men would begin at once to make all property public, available for the use and enjoyment of all the people. The early disciples of Jesus understood this and their religion required them to bring all their possessions and lay them at the feet of their leaders, to be distributed and used according to the needs of all. Paul's companion Barnabas gave his field. Ananias and Sapphira sold their land and brought part of the price to Peter but held back part of it. They had not overcome the fear of future lack and had not put their faith fully in the teaching and promises of the Master.

When we have recognized the truth of the omnipresence of God as substance and supply for every need, there will be no occasion for holding back part as Ananias and Sapphira did. We cannot hoard money in its material phase without breaking the law, which is that we have all the substance necessary for our supply. We ask the Lord for our "daily" bread and expect to have it but we do not get an accumulation that will spoil on our hands or that will deny the proper supply to any other man. The metaphysical idea of this part of the Lord's Prayer is "Give us this day the substance of tomorrow's bread." We ask not for bread but for the substance that Spirit arranges to manifest as bread, clothing, shelter, or the supply for any need we may have.

Substance in the form of money is given to us for constructive uses. It is given for use and to meet an immediate need, not to be hoarded away or be foolishly wasted. When you have found freedom from the binding thought of hoarding money, do not go to the opposite extreme of extravagant spending. Money is to be used, not abused. It is good to keep one's obligations paid. It is good to have some money on hand for good uses, such as hospitality, education, for developing industries that will contribute to the good of numbers of people, for the furtherance of spiritual work, for helping others to build useful and constructive lives, and for many other purposes and activities. But in such conservation of money one should keep ever in mind the necessity of a constructive motive back of the action. Money accumulated for a definite and definitely constructive purpose is quite a different thing from money hoarded with the fearful thought of a "rainy day" or a prolonged season of lack and suffering. Money saved for "rainy days" is always used for just that, for fear attracts that which is feared unfailingly. "The thing which I fear cometh upon me."

Money saved as "an opportunity fund" brings an increase of good, but money hoarded from fear as a motive or with any miserly thought in mind cannot possibly bring any blessing. Those who hold the thought of accumulation so dominant in the world today are inviting trouble and even disaster, because right along with this thought goes a strong affirmation of the fear of loss of riches. Their actions bespeak fear, and the loss they dread is certain to be manifested sooner or later. The worldly idea of prosperity is based on the wrong idea of supply. One may have the right idea about the source of riches as spiritual and yet have the wrong idea about the constancy of supply as an ever-present, freely flowing spiritual substance. God does not clothe the lilies in a moment and then leave them to the mercy of lack; He gives them the continuous supply necessary to their growth. We can rest assured that He will much more clothe us and keep us clothed from day to day according to our need. When we doubt this and place our dependence on stored-up money instead, we shut off the stream of divine supply. Then when our little accumulation is spent, stolen, or lost, we are like the prodigal son and we begin to be in want.

Jesus did not own a foot of land. Yet never did He lack for anything needed. Without laying up treasures on earth He was rich in His consciousness of the treasures of heaven within Himself, treasures ready to be manifested in the outer whenever He needed them.

We know perfectly well that sooner or later we shall have to let go of our earthly possessions. Does this bring the thought of death and of leaving the world behind? Then it shows what a powerful hold this race belief of worldly wealth has taken in your mind. Men can think of letting go of their material possessions only in connection with death. They seem to prefer death to giving up their idea of wealth. When they make such a choice they decree what shall come to pass for them. That is why it is hard for a "rich man" to enter into the kingdom of heaven. He has laid up treasures on earth and not enough in heaven. He has not made it possible for his mind to lay hold of the positive pole of wealth, the true idea of wealth. He is holding to the negative side of the wealth idea, and that side is always changing. Material things pass away unless they are firmly connected with the unchanging, positive Source.

True riches and real prosperity are in the understanding that there is an omnipresent substance from which all things come and that by the action of our mind we can unify ourselves with that substance so that the manifestations that come from it will be in line with our desires and needs. Instead of realizing the inexhaustible, eternal, and omnipresent nature of that substance, we have limited it in our thought. We have thought that there is only about so much of it and that we had better hurry to get our share. We have thought that we must be careful how we spend it and put some of it away for a time when there won't be any more. In building up this consciousness of a limited supply we have concluded that it is necessary to be economical and more and more saving. We begin to pinch in our mind, and then our money becomes pinched, for as we think in our mind, so we manifest in our affairs. This attitude pinches the channel through which our substance comes to manifestation and slows down the even flow of our supply. Then comes depression, hard times, shortage, and we wonder why, looking for some way to lay the blame on the government, or on war, or on industry, or even on the Lord, but never by any chance do we put the blame where it belongs: on ourselves.

The "pinching attitude" of mind does even worse than bring people into want. If people would relax in mind, they would loosen up the nerves and muscles of the body. They must learn the cause of their strained, pinching mental attitude and let go of that first. Then the relief of the outer condition will become manifest as the condition itself did.

Nearly all of us have been brought up in the belief that economy is an important thing, even a virtue. We should save our money and have a bank account. Saving money is the recipe for success given by many of our wealthy men. It is not a bad idea. There must be money available in banks to carry on business and industry. By having a bank account we contribute to the welfare of the community, if we have the right idea; which is that the Lord is our banker.

The word miser is from the Latin root from which also comes "miserable." It describes the condition of those who love and hoard money, lands, or other material things. The stories that are told about misers are almost beyond credence, but nearly every day the press recounts the story of the pitiable straits to which misers have reduced themselves in order to add to their riches. They sometimes starve themselves to add a few dollars or even a few pennies to their hoarded store. The papers recently carried an item about a miser in New York worth eleven million dollars. He goes from office to office in one of his great office buildings and picks up the waste paper from the baskets, which he sells for a few cents. Another almost as wealthy will not buy an overcoat but keeps his body warm by pinning newspapers under his house coat. Such men are not only themselves miserable but they make miserable all those around them. A New York paper tells of a miser worth millions when he died. Once burglars broke into his home, but they succeeded in getting out again without losing anything.

You do not need to lay up treasures for the future when you know that the law of omnipresent good is providing for you from within. As you evolve into this inner law of mind, you draw to yourself more and more of the good things of life.

In your mind see plenty everywhere. Yes, it is hard sometimes to overcome the thought that there is not enough, for it is an insidious thought that has been in consciousness for a long time. But it can be done. It has been done and is being done by others. The prosperity law is not a theory but a demonstrated fact, as thousands can testify. Now is the time to open your mind and to see plenty. As you do so you will find that there is an increase in your supply. Deny out of mind every thought of lack and affirm the abundance of all good. The infinite substance that infinite Mind has given to you is all about you now, but you must lay hold of it. It is like the air, but you must breathe the air to get it. It is yours for the taking, but you must take it. You should cultivate this wonderful power of the mind to know that everything is bountiful and this power to lay hold of invisible substance in the mind and by faith bring it forth into manifestation. Know with Job that we have as much now, in reality and in Truth, as we ever had. There is no shortage, lack, or depression with God.

Do not be fearful, regardless of how outer appearances may affect others. Keep your head when all about you are losing theirs. Refuse to load up your mind with the old material thoughts of economy to the point of denial of what you really need. Eliminate the old limiting ideas. Assert your freedom and your faith as a child of God. Do not spend foolishly or save foolishly. The farmer does not throw away his wheat when he sows a field. He knows how much he must sow per acre and does not stint, for he knows that a stinted sowing will bring a stinted harvest. He sows bountifully but not extravagantly and he reaps bountifully as he has sown. "Whatsoever a man soweth, that shall he also reap." "He that soweth sparingly shall reap also sparingly; and he that soweth bountifully shall reap also bountifully."

We cannot help but see that apparent lack and hard times are the result of states of mind. We have such things in the manifest world because men have not squared their action with divine Principle. They have not used spiritual judgment. When they invest in stocks and property, they get the opinions of other men, sometimes those who call themselves experts. Then comes the crash, and even the experts prove how little they understand the real laws of wealth. We can go to an expert who really knows the law because He ordained it in the first place. And He is not far away, but right within ourselves. We can go within and meditate on these things in the silence, and the Lord will direct our personal finances. He will show us just how to get the most and give the most with our money and He will see to it that we have the supply that we need so that we may not be in want of anything needful to our good. This may not mean riches piled up or "saved for a rainy day," but it will insure our supply for today, the only day there is in Truth.

As we continue to grow in the consciousness of God as omnipresent life and substance we no longer have to put our trust in accumulations of money or other goods. We are sure that each day's need will be met, and we do not deprive ourselves of today's enjoyment and peace in order to provide for some future and wholly imaginary need. In this consciousness our life becomes divinely ordered, and there is a balance in supply and finances as in everything else. We do not deprive ourselves of what we need today; neither do we waste our substance in foolish ways nor deplete it uselessly. We do not expect or prepare for adversity of any kind, for to do so is not only to invite it but to show a doubt of God and all His promises. Many people bear burdens and deny themselves sufficient for their present needs in order to prepare for dark days that never come. When we look back over the past we find that most of our fears were groundless, and most of the things we dreaded so much never happened. However the things we prepared for probably did happen and found us not fully prepared even after all our efforts in that direction. This should enable us to trust God now and rest in the positive assurance that He will supply every need as it arises.

Things are never so bad as you think. Never allow yourself to be burdened with the thought that you are having a hard time. You do not want a soul structure of that kind and should not build it with those thoughts. You are living in a new age. Yesterday is gone forever; today is here forever. Something grander for man is now unfolding. Put yourself in line with the progress of thought in the new age and go forward.

Lesson 12
Overcoming the Thought of Lack

The kingdom of heaven is like unto a net, that was cast into the sea, and gathered of every kind; which, when it was filled, they drew up on the beach; and they sat down, and gathered the good into vessels, but the bad they cast away."

The mind of man is like the net catching every kind of idea, and it is man's privilege and duty under the divine law to separate those that are good from those which are not good. In this process the currents of unselfish, spiritual love flowing through the soul act as great eliminators, freeing the consciousness of thoughts of hate, lack, and poverty, and giving the substance of Spirit free access into the consciousness and affairs.

In another parable Jesus explained the same process as a separation of the sheep from the goats. When this divine current of love and spiritual understanding begins its work, we must make this separation. We put the sheep, the good and obedient and profitable thoughts, on the right, and we put the goats, the stubborn, selfish, useless thoughts, on the left. Each must handle his own thoughts and overcome them by aligning them with the harmony and order of the divine thought. There is an infinite, omnipresent wisdom within us that will deal with these thoughts and guide us in making the discrimination between the right and the wrong when we trust ourselves fully to its intelligence. We can establish a connection between the conscious mind and the superconscious mind within us by meditation, by silence, and by speaking the word.

The superconscious mind within you discriminates among the kinds of food you assimilate, controls your digestion, your breathing, and the beating of your heart. It "doeth all things well," and it will help you do this important work of directing you in the thoughts you should hold and the ones you should cast out. As you develop this mind within yourself you will find that you can gradually turn over more and more of your affairs to its perfect discrimination. Nothing is too great for it to accomplish, nor is anything too trivial for it to handle with perfection and dispatch. This mind of the Spirit will guide you in perfect ways, even in the minute details of your life, if you will let it do so. But you must will to do its will and trust it in all your ways. It will lead you unfailingly into health, happiness, and prosperity, as it has done and is doing for thousands, if and when you follow it.

It is just as necessary that one should let go of old thoughts and conditions after they have served their purpose as it is that one should lay hold of new ideas and create new conditions to meet one's requirements. In fact we cannot lay hold of the new ideas and make the new conditions until we have made room for them by eliminating the old. If we feel that we cannot part with the goats, we shall have to do with fewer sheep. If we insist on filling the vessels with the bad fish, we shall have to do without the good. We are learning that thoughts are things and occupy "space" in mind. We cannot have new or better ones in a place already crowded with old, weak, inefficient thoughts. A mental house cleaning is even more necessary than a material one, for the without is but a reflection of the within. Clean the inside of the platter, where the food is kept as well as the outside that people see, taught Jesus.

Old thoughts must be denied and the mind cleansed in preparation before the affirmative Christ consciousness can come in. Our mind and even our body is loaded with error thoughts. Every cell is clothed with thought: every cell has a mind of its own. By the use of denial we break through the outer crust, the material thought that has enveloped the cells, and get down into the substance and the life within them. Then we make contact with that substance and life which our denials have exposed, and by it express the positive, constructive side of the law. When we consistently deny the limitations of the material, we begin to reveal the spiritual law that waits within ourselves to be fulfilled. When this law is revealed to our consciousness, we begin to use it to demonstrate all things that are good. That is the state of consciousness that Jesus had, the Christ consciousness.

Every man has a definite work to do in the carrying forward of the divine law of spiritual evolution. The law is set into action by our thinking and is continually supported by our thought as it develops our soul. Within us are the great potentialities of Spirit that, put into action, enable us to be, do, or have anything we will. Science tells us that each of us has enough energy within himself to run a universe, if we knew how to release and control it. We do this releasing by a process of letting go and taking hold: letting go of the old or that which has done its part and is no longer useful, and taking hold of the new ideas and inspirations that come from the superconscious mind. Jesus told Peter that what he should bind in earth would be bound in heaven and what he should loose in earth would be loosed in heaven. He was not talking about a geographical earth or a definite place in the skies called heaven. He was explaining to Peter the law of mind. The conscious mind is but the negative pole of a very positive realm of thought. That positive realm of thought, Jesus called "the kingdom of the heavens." It is not a place at all but is the free activity of the superconscious mind of man. Whatever we bind or limit in earth, in the conscious mind, shall be bound or limited in the ideal or heavenly realm, and whatever we loose and set free in the conscious mind (earth) shall be loosed and set free in the ideal, the heavenly. In other words, whatever you affirm or deny in your conscious mind determines the character of the supermind activities. All power is given unto you both in heaven and in earth through your thought.

We must carefully choose what thoughts we are going to loose in the mind and what thoughts we are going to bind, for they will come into manifestation in our affairs. "As he [man] thinketh within himself, so is he" and "whatsoever a man soweth [in the mind], that shall he also reap [in the manifestation]." We must loose all thoughts of lack and insufficiency in the mind and let them go, just as Jesus commanded be done with the wrappings that held Lazarus: "Loose him, and let him go." Loose all thoughts of lack and lay hold of thoughts of plenty. See the abundance of all good things, prepared for you and for all of us from the foundation of the world. We live in a very sea of inexhaustible substance, ready to come into manifestation when molded by our thought.

Some persons are like fish in the sea, saying, "Where is the water?" in the presence of spiritual abundance they cry, "Where will I get the money? How will I pay my bills? Will we have food or clothes or the necessities?" Plenty is here, all around, and when you have opened the eyes of Spirit in yourself, you will see it and rejoice.

We mold omnipresent substance with our mind and make from it all the things that our mind conceives. If we conceive lack and poverty we mold that. If we visualize with a bountiful eye we mold plenty from the ever-present substance. There is perhaps no step in spiritual unfoldment more important than the one we are taking here. We must learn to let go, to give up, to make room for the things we have prayed for and desired. This is called renunciation or elimination, sacrifice it may even seem to some people. It is simply the giving up and casting away of old thoughts that have put us where we are, and putting in their place new ideas that promise to improve our condition. If the new ideas fail to keep this promise, we cast them away in their turn for others, confident that we shall eventually find the right ideas that will bring that which we desire. We always want something better than we have. It is the urge of progress, of development and growth. As children outgrow their clothes we outgrow our ideals and ambitions, broadening our horizon of life as we advance. There must be a constant elimination of the old to keep pace with this growth. When we cling to the old ideals we hinder our advance or stop it altogether.

Metaphysicians speak of this eliminative work as denial. Denial usually comes first. It sweeps out the debris and makes room for the new tenant that is brought into the mind by the affirmation. It would not be wise to eliminate the old thoughts unless you knew that there are higher and better ones to take their place. But we need not fear this, because we know the divine truth that God is the source of all good and that all good things can be ours through the love and grace of Jesus Christ.

None of us has attained that supreme place in consciousness where he wholly gives up the material man and lives in the Spirit, as Jesus did, but we have a concept of such a life and His example showing that it can be attained. We shall attain it when we escape the mortal. This does not mean that we must die to get free from mortality, for mortality is but a state of consciousness. We die daily and are reborn by the process of eliminating the thought that we are material and replacing it with the truth that we are spiritual. One of the great discoveries of modern science is that every atom in this so-called material universe has within it superabundant life elements. God is life and Spirit, and He is in every atom. We release this spiritual life quality by denying the crust of materiality that surrounds the cells and affirming that they are Spirit and life. This is the new birth, which takes place first as a conception in the mind, followed by an outworking in body and affairs. We all want better financial conditions. Here is the way to obtain them: Deny the old thoughts of lack of money and affirm the new thought of spiritual abundance everywhere manifest.

Every lesson of Scripture illustrates some phase of mental action and can be applied to each individual life according to the need that is most pressing at the time of its perception. If you do not look for the mental lesson when reading Scripture, you get but the mere outer shell of Truth. If however you have the proper understanding of the characters in the narrative, knowing that they represent ideas in your own mind, you can follow them in their various movements and find the way to solve all the problems of your life. This does not mean that a study of the written Scriptures will itself solve your problems unless you come into the apprehension of the real Scriptures, the Bible of the ages, the Book of Life within your own consciousness. But a study of the outer symbols as given in the written Scriptures can and should lead you into the understanding of the Truth of Being.

In every person we find the conflicting ideas represented by the Children of Israel and the Philistines. They are pitted against each other in a conflict that goes on night and day. We call these warring thoughts Truth and error. When we are awakened spiritually we stand on the side of Truth, knowing that Truth thoughts are the chosen of the Lord, the Children of Israel. But the error thoughts sometimes seem so real and so formidable that we quake and cringe with fear in their presence.

We know that Truth will eventually prevail, but we put the victory off somewhere in the future and say that the error is so large and strong that we cannot cope with it now--we will wait until we have gathered more strength. Then we need to stand still and affirm the salvation of the Lord.

Ideas are not all of the same importance. Some are large and strong; some are small and weak. There are aggressive, dominating ideas that parade themselves, and brag about their power, and with threats of disaster keep us frightened into submission to their wicked reign. These domineering ideas of error have one argument that they always use to impress us, that of the fear of results if we should dare to come out and meet them in open opposition. This fear of opposing ideas, even when we know them to be wrong, seems to be woven into our very mental fabric. This fear is symbolized by the spear of Goliath which, as the story relates, "was like a weaver's beam."

What is the most fearful thought in the minds of men today? Is it not the power of money? Is not mammon the greatest Philistine, the Goliath in your consciousness? It is the same whether you are siding with the Philistines and are successful in your finances from a material viewpoint, or whether you are with the Israelites and tremble in your poverty. The daily appearance of this giant Goliath, the power of money, is something greatly feared. Neither the Philistines nor the Israelites are in possession of the Promised Land, neither side at peace or happy in any security, so long as this domineering giant parades his strength and shouts his boasts. This error idea claims he is stronger than the Lord of Israel. He must be killed before all the other error thoughts will be driven out of your consciousness and you can come into the consciousness of plenty, the Promised Land of milk and honey.

The whole world today trembles before this giant error idea, the belief that money is the ruling power. The nations of the world are under this dominion because men think that money is power. The rich and the poor alike are slaves to the idea. Kings and great men of the earth bow and cringe before the money kings. This is because man has given this power to money by his erroneous thinking. He has made the golden calf and now he falls before it in worship. Instead of making it his servant he has called it master and become its slave. The rule of this mad giant has been disastrous, and the end of it is rapidly approaching.

The first step in getting your mind free from this giant bugaboo is to get a clear perception of your right as a child of God. You know that you should put no other gods or powers before the true God. You know also that you should not be under the dominion of anything in the heavens above or the earth beneath, for you have been given dominion over all. You will never find a better time to come into the realization of the truth of who and what you are and what your rights are. Never was a more propitious time to seek a new and better state of consciousness. If you are in fear of the boasting Philistine giant, as so many around you are, begin now to seek a way, as did David, to give his "flesh unto the birds of the heavens." There is a way, a righteous way, that cannot fail, and it is your duty to find it. Follow each step of the way that is symbolically and beautifully set forth in the 17th chapter of I Samuel.

The name David means "the Lord's beloved," and David represents your righteous perception of your privileges as the child of God. You are not a slave to anything or to anybody in the universe. The threat of this Goliath, the power of money, holds no terrors for you in this consciousness. You have a smooth perception of Truth and you sling it straight to the center of his carnal thinking, his forehead. The weight of his shield and his armor does not intimidate you, for you see them for what they are, empty and meaningless show, vulnerable in many places to the true ideas with which you are armed.

Even the most ardent defenders of the money power will admit that it is a tyrant and that they would not have it rule their world if they could help it. It nearly always destroys its friends in the end. Any man who becomes a slave to money is eventually crushed by it. On the other side are whole armies of righteous people, Christians, who like the army of Israel think that this giant cannot be overcome. They are waiting for reinforcements, something larger and stronger in a physical way, with which to overcome this enemy. They forget that "the battle is Jehovah's."

Do you cringe before this giant when he comes out daily to impress you with his boastings and threats? It does not have to be so. You need not continue to fear. There is a little idea in your mind that can slay him. You perhaps have not considered this little idea of much importance. Perhaps you have kept it off on a lonely mountainside of your spiritual consciousness, herding the sheep, which are your innocent thoughts. Now let this David come forth, this perception of your rightful place in Divine Mind. Get a clear idea of where you really belong in creation and what your privileges are. Do you think for a moment that God has so ordained that men cannot escape from the terrible servitude of hard conditions? Of course not. That would be injustice, and God is above all just.

It is your privilege to step out at any time and accept the challenge of this boaster. The Lord has been with you in the slaying of the fear of sin and sickness (the bear and the lion), and He will still be with you in slaying the fear of poverty, which Goliath symbolizes. "The battle is Jehovah's," and He is with us to deliver us "out of the hand of the Philistines."

The weapons of the Lord's man are not carnal. He does not wage war after the manner of the world. He does not use armor of steel or brass, the protection of selfishness and the weapons of oppression. He goes forth in the simplicity of justice, knowing that his innocence is his defense. He uses only his shepherd's sling and smooth stones, words of Truth. This is the will and the words of Truth that it sends forth. They are disdained by the Philistines and many people laugh at the idea of using words to overcome conditions. But words do their work, the work whereto they are sent, and the great mass of materiality goes down before their sure aim.

We know that money was made for man and not man for money. No man needs to be a slave to his brother man or cringe before him to obtain money, which is the servant of all alike. We are not bound to the wheel of work, of ceaseless toil day after day, in order to appease the god of mammon on his own terms. We are children of the living God, who as a loving Father is right here in our midst, where we may claim Him as our support and our resource on such conditions as He lovingly reveals when we have acknowledged Him and denied mammon. This day has Jehovah delivered this proud Philistine into our hands, and the victory is ours. Praise God.

The five smooth stones chosen by David from the brook represent five irrefutable statements of Truth. These statements sent forth from a mind confident of itself, its cause, and its spiritual strength will crush the forehead of Goliath, error's giant. The statements are the following:

I am the beloved of the Lord. He is with me in all my righteous words, and they do accomplish that whereto I send them forth.

My cause is just, for it is my divine right to be supplied with all things whatsoever that the Father has placed at the disposal of His children.

I dissolve in my own mind and the minds of all others any thought that my own can be withheld from me. What is mine comes to me by the sure law of God, and in my clear perception of Truth I welcome it.

I am not fearful of poverty, and I am under obligations to no one. My opulent Father has poured out to me all resources, and I am a mighty channel of abundance.

I selfishly own nothing, yet all things in existence are mine to use and in divine wisdom to bestow upon others.

Do not hold yourself in poverty by the fear of lack and by practicing a pinching economy. If you believe that all that the Father has is yours, then there is surely no reason for skimping. Nothing will so broaden your mind and your world as the realization that all is yours. When you realize the boundlessness of your spiritual inheritance, nothing shall be lacking in all your world. See with the bountiful eye; for "he that hath a bountiful eye shall be blessed." This passage states an exact law, the law of increase.

Religious leaders in the past have spread the belief that it is a Christian duty to be poor and that poverty is a virtue. This is by no means the doctrine of Jesus. He accepted the proposition fully, without reservation or qualification, that God is our resource and that the Father has provided all things for His children. He is often described as being poor, without a place to lay His head, yet He had a parental home at Nazareth and was welcomed gladly into the homes of both the rich and the poor all over Palestine. He dressed as a rabbi, and His clothing was so rich and valuable that the Roman soldiers coveted the seamless robe He wore and cast lots for it. He found abundance in the kingdom of God where everything needful becomes manifest not through hard labor but through the realization of Truth.

Jesus seldom had need for money, because He went back of money to the idea it represents and dealt with money in the idea realm. Our government is back of all our paper dollars, else they would have no value. God is back of every material symbol, and it is in God rather than in the symbol that we should put our faith. He is back of our call for food and raiment and everything that we could need or desire. Jesus says all we need do is ask in faith and in His name, believing that we receive, and we shall have. And we should not hesitate to ask largely, for God can give much as easily as He can give a little.

Question Helps

Lesson 1. Spiritual Substance, The Fundamental Basis Of The Universe

1. What is Divine Mind?

2. What is man, and how is he connected with divine ideas?

3. What great change in methods of production and distribution seems about to be made? How will it affect our prosperity?

4. What is the ether of science and metaphysics? To what extent has man drawn on it, and what are its possibilities?

5. What did Jesus demonstrate regarding the kingdom of the ether?

6. What is the source of all material, according to science? According to Jesus?

7. What is the simplest and surest way to lay hold of substance?

8. Explain from this viewpoint how substance can never be depleted.

9. Why does God give to just and unjust alike, to all equally?

10. How does this truth of the ether help us better to understand the nature of God as pure being or Spirit?

11. What is symbolized by gold and silver? Why are they precious?

12. What is the threefold activity through which substance must go on its way to becoming manifest as material?

13. If substance is omnipresent and man can control its manifestation, why does man suffer from lack and limitation?

14. Explain the teaching of Jesus that it is hard for a rich man to enter the kingdom of the heavens.

15. What is meant by "property rights" and the right to wealth? What error is implied in this doctrine? To whom do ideas belong?

16. What are some of the "great possessions" that must be unloaded before we can enter the kingdom of consciousness?

17. After recognizing the existence, potentiality, and availability of universal substance, what is the next step in demonstration?

18. Can the kingdom be found by one with selfish motives? Why should we desire healing and prosperity?

19. What is the prosperity consciousness? Give examples. How can it be cultivated wisely?

20. What will be the social and economic results of a widespread prosperity consciousness in the whole race?

Lesson 2. Spiritual Mind, The Omnipresent Directive Principle Of Prosperity

1. Why are ideas the most important things in life?

2. What is desire in origin, purpose, and result?

3. What is the difference between "is-ness" and "existence"?

4. What is the difference between "being" and "appearance"?

5. What is the relation of figures to the problem they help solve? How does this illustrate spiritual reality and material phenomena?

6. What is implied in the fact that man can conceive of an ideal world?

7. Should we deny the existence of material things? Can we do so successfully? What should we deny about the things of the outer?

8. What is the "I AM identity"? How does it differ from Divine Mind?

9. Why is spiritual understanding important? How is it gained?

10. What connection exists between ideas in Divine Mind?

11. What divine idea is back of riches? What ideas are the "parents" of this idea? How can this knowledge help us in demonstrating?

12. Are all men equally entitled to wealth? What ideas should accompany the acquiring, using, and spending of wealth?

13. Do we expect God to give us actual loaves of bread when we pray the Lord's Prayer? What does He give instead of material things?

14. Why do people have dreams? Do dreams help men with their problems?

15. What is the value of relaxation and the silence when we seek God's gifts?

16. What relation has prosperity demonstration to the kingdom-of-the-heavens consciousness?

17. What is the physical, psychological, and spiritual reason for preparing the way for the prosperity demonstration?

18. What does the parable of the lilies teach us about substance?

19. What effect does the attitude of thanksgiving and praise have upon our prosperity?

20. Why is asking in the name of Jesus Christ more effective than any other prayer?

Lesson 3. Faith In The Invisible Substance, The Key To Demonstration

1. What is the starting point in building a prosperity consciousness?

2. What is the relation between faith and substance?

3. What does it mean to "have" faith?

4. What is a "seeking" faith? For what does it seek?

5. Explain how doubt retards manifestation.

6. What is the difference between the conception of John the Baptist and that of Jesus?

7. How do love and understanding assist faith in its accomplishments?

8. Are difficult experiences necessary in life? Why do we have them?

9. Show that it is sinful to think and talk hard times, lack, and other limitations.

10. Explain the symbology of the five loaves and two fishes.

11. How does fear produce a stagnation in financial circulation? How does confidence or faith restore normal conditions?

12. How can we go into "the upper room" to wait for the power from on high?

13. How does your mind create? Are its creations always real?

14. Why should our faith be in Spirit rather than in material things?

15. Show how faith is essential to success in the professions, in manufacturing, in sales, and in other lines of activity.

16. What do the Bible characters represent to us today? What Bible personage represents faith?

17. Is science antagonistic to religion or helpful to its cause?

18. What is the relation between material and substance?

19. Is there any lack of anything anywhere? What are we to overcome?

20. What affirmations help most to banish fear and abide in the consciousness of plenty?

Lesson 4. Man, The Inlet And Outlet Of Divine Mind

1. What is meant by Principle as applied to prosperity?

2. How do we establish a consciousness of Principle as related to us?

3. How can the study of Truth make one happier, healthier, more beautiful, more prosperous?

4. What is a miracle? Is prosperity miraculous?

5. How are the keepers of divine law rewarded, and its breakers punished?

6. What are the legislative, judicial, and executive phases of the divine law?

7. What is the first rule of the divine law?

8. What is the effect of thinking and speaking of everything as good?

9. Is there any virtue in poverty?

10. What is meant by the "far country," and what is the homeland of the prodigal son?

11. What is the psychological and spiritual effect of old clothes?

12. What is symbolized by the putting on of new shoes.

13. How is true substance wasted, and what is the connection between waste and want?

14. What is our best insurance of financial security?

15. How does the law "Seek and ye shall find" apply to prosperity?

16. Should one who works harder or has more ability receive a greater reward than another?

17. What power has love in helping one to demonstrate prosperity?

18. How does the subconscious mind help or hinder in demonstration?

19. What form does God's answer to prayer take? How do we know when a prayer is answered?

20. Must one be morally worthy to become prosperous?

Lesson 5. The Law That Governs The Manifestation Of Supply

1. What in our consciousness is represented by Moses? By Joshua? By Jesus?

2. What is the metaphysical significance of eating? How do we break bread in the four-dimensional world?

3. What retards manifestation when we work to attain the consciousness of abundance?

4. What do we mean by the "one law"? How may we know it? How keep it?

5. Explain how our ability to use wealth wisely to a large degree determines our prosperity.

6. Is it necessary to beseech God for prosperity? To ask? To thank?

7. How do we look or go "within"?

8. Compare the sense mind with the spiritual mind and show how true prosperity depends on the latter.

9. What is the light theory of matter formation, and how does it agree with the teaching of the New Testament?

10. Where and what is heaven? How is the soul formed?

11. How did King Solomon demonstrate great prosperity?

12. What is meant by "laying hold of" the substance?

13. How did Jesus develop His consciousness of omnipresent substance and what did that consciousness ultimately do for Him?

14. What is the true interpretation of "rich man" in the famous parable of the camel and the needle's eye?

15. How do we constantly "turn stones into bread," and what results?

16. What does the parable of the talents teach us regarding prosperity?

17. What six steps necessary to manifestation may be discerned in the story of creation?

18. What do we contribute to the world by raising our own consciousness to the prosperity level?

19. Analyze and explain the statement "I trust Thy universal law of prosperity in all my affairs."

Lesson 6. Wealth Of Mind Expresses Itself In Riches

1. What is prosperity?

2. Explain the prosperity law that Jesus gave.

3. What relation has a prosperity consciousness to wealth in the outer?

4. What is the "sin of riches"?

5. What causes crop failure and famine in some countries?

6. Why did Jesus carry no money and own no property?

7. What is the only thing that can satisfy human longing, and where is it found?

8. Is the law of prosperity limited to thought? What else is needful?

9. Over what is man given dominion by his Creator?

10. How can man master his fear of financial lack?

11. Why are prosperity prayers sometimes unanswered?

12. What is the true idea of God, and how does man give it form?

13. What causes "depressions" in the affairs of men and nations?

14. What part does self-control in the matter of sensation play in prosperity demonstration?

15. What is the relative importance of denial and affirmation in the demonstration of prosperity?

16. Who are the real producers of wealth in the nation?

17. What is the law of increase as applied to Mind substance?

18. How should we prepare for an increased prosperity?

19. Should we be specific and definite in our prayers for increase?

20. Write a prosperity affirmation of your own embodying the four essential steps of recognition, love, faith, and praise.

Lesson 7. God Has Provided Prosperity For Every Home

1. Of what great spiritual power is the home the symbol?

2. What has the "atmosphere" of a home to do with its prosperity? How may an atmosphere of worry and fear be changed?

3. Explain the importance of speaking true words in the home.

4. Aside from the feeling of religious duty, why should we be thankful for what we have and express our thanks often?

5. Is it good policy to condemn the furnishings in the home or to be apologetic about them?

6. Should our homes be ostentatious and rich looking to attract prosperity?

7. Why should we be individual in furnishing the home rather than following the "accepted" or "in-the-mode" style only?

8. How will a deep and sincere love for God attract prosperity?

9. Why must there be love and understanding between members of the family to insure a prosperous home?

10. Explain the law of "Love thy neighbor" as applied to home prosperity.

11. What is God's will for the home, and how does the home express it?

12. Explain how trying to live and do as others live and do may hold back our prosperity demonstration.

13. How can we use our will to help the demonstration of home prosperity?

14. Why should the individual express his own ideas in order to demonstrate?

15. Where and how is the prosperity demonstration started?

16. Do we have any personal claim on God's substance?

17. How do we "pour" substance into the "empty" places of the home?

18. Why is it necessary to have determination in order to demonstrate?

19. Does the possession of material things give satisfaction?

Lesson 8. God Will Pay Your Debts

1. What law of mind is observed in true forgiveness?

2. Why should we trust rather than distrust people?

3. Is there any such thing as debt in Truth? Why?

4. Where must we start in forgiving our debtors and creditors?

5. How can we forgive ourselves for holding others in our debt?

6. What is the only sure way of getting out and staying out of debt?

7. Explain forgiveness as a good method of bill collecting.

8. How does God forgive our debts? How does His love pay our debts?

9. How do debt and worry about debt affect health? What is the remedy?

10. Does God have a place in modern business?

11. What are the merits and the dangers of installment buying?

12. What is the importance of paying all obligations promptly?

13. What kind of thoughts should one hold toward creditors? Debtors?

14. What dominant belief has caused world depression, and how must it be overcome? What is our part in its overcoming?

15. Is the credit system responsible for widespread debt?

16. Does our faith in supply justify us in assuming obligations and trusting that we shall be able to pay when the time comes?

17. What is the value of prayer in gaining freedom from debt?

Lesson 9. Tithing, The Road To Prosperity

1. What is a "tithe," and how was tithing started?

2. What benefits accrue to the tither, according to the promises of the Bible?

3. Should one regard one's tithe as an investment that pays rewards?

4. In what way is giving a divine grace?

5. What was the practical plan that Paul suggested to the Corinthians?

6. What effect does a willing and cheerful spirit have

on the giver, the gift, and the receiver?

7. How can faith be exercised in giving?

8. How should wisdom be employed in giving?

9. How can one who is puzzled about giving--as regards how much, when, and where--be helped by the decision to tithe?

10. What should tithing mean to the farmer? Businessman? Professional man? Mechanic? Laborer?

11. How does tithing help fulfill "the first and greatest commandment" about loving God and the neighbor?

12. Aside from Bible promises, do we have direct evidence that tithing increases prosperity? Cite instances.

13. Should the tithe have a definite place in the personal or family budget? Should we keep a record of our giving, as we do of other disbursements?

14. Why is the regular tithe, though it may be small, better than the occasional giving of a larger gift in a lump sum?

15. What is the psychological basis and effect of tithing?

16. What attitude should one assume toward a seemingly delayed demonstration?

17. Should we look for our good to come through the channel of those to whom we give or serve?

18. Why is it better to give without thought or expectation of return?

19. What must we do about receiving what God has and desires for us?

20. Discuss giving as a form of affirmation.

Lesson 10. Right Giving, The Key To Abundant Receiving

1. In what ways is the religion of Jesus applicable to the problems of daily living?

2. State briefly the law of giving and receiving that Jesus taught.

3. Why has the teaching of Jesus not been more effective in changing conditions in the world and in individual life?

4. Why is economic reform so much needed at the present time?

5. Can any effective reform be based on the material phase of the economic problem? Why?

6. Why do men who direct finance and business fail to seek any advice or assistance from the church?

7. Why is individual reform necessary before national or world changes can be made?

8. What do metaphysical teaching and study contribute toward world betterment?

9. How does the desire for the accumulation of money and goods affect the finer nature and sensibilities of people?

10. Does avarice or greed have any effect on the health of men?

11. What is the chief cause of stagnation in money circulation and its attendant evils?

12. What rule did Jesus give us for freeing ourselves from financial lack?

13. Is the method practiced by the early Christians practicable in the world as it is today?

14. What substitute is now being advocated for the commercial standard of payments for goods and service?

15. What is the divine law of equilibrium? Why does it not seem to operate in financial matters?

16. Is there any direct connection between poverty and ill-health? How may this problem be approached? Is there a problem at the other extreme--great wealth? How may it be solved?

17. What is meant by the "race consciousness"? Can we escape its effects? How can we help to change it for the better?

18. What should we do about saving money for the future?

19. What attitude should we take toward charity?

Lesson 11. Laying Up Treasures

1. What is the law of conservation as applied to spiritual substance?

2. What is the difference between hoarding and conserving?

3. Is accumulation of substance necessary to demonstration?

4. How is spiritual substance accumulated? How dissipated?

5. What is the true objective of man's life?

6. Explain why character development must be a part of our study of the demonstration of prosperity.

7. Is the ambition for wealth commendable or reprehensible?

8. Of what is gold the symbol, metaphysically understood?

9. What is the deceitfulness of riches? What is money?

10. Does great wealth bring happiness? Does extreme poverty make one any better than the rich? What is the truth about riches?

11. What is the only true deed or title to possessions?

12. What do we want when we ask for "our daily bread"?

13. How does the hoarding of money injure society?

14. Should we prepare for that "rainy day" by saving part of our money?

15. Was Jesus poor? Was He ever in want? What does it mean to turn stones into bread?

16. Explain the meaning of the rich man and the eye of the needle.

17. What are some of the financial and bodily results of the pinching attitude toward money?

18. What attitude toward hard times and lack is most helpful to us?

19. How shall we learn to get the most and give the most with the means we have at our disposal?

Lesson 12. Overcoming The Thought Of Lack

1. Why must we constantly examine our thoughts and separate them?

2. By what standard do we judge our thoughts?

3. What is the work of the superconscious mind in the body?

4. How may we use this superconscious mind in our outer affairs?

5. What is the importance of thought elimination or mental clean-up?

6. How do we go about this work of eliminating error thoughts?

7. What further benefits accrue from the use of denial words?

8. Where or what is the "kingdom of the heavens," and what is it like?

9. What is meant by "loosing in heaven"? What should we loose and what should we guard against loosing?

10. Where and what is "substance"? How do we contact it?

11. What is the result of clinging to past ideas and methods?

12. What is mortality? Do we escape it by dying? How otherwise?

13. Explain the new birth and the relation of denial to it.

14. What do Bible characters mean to us? What can they do for us?

15. What do David and Goliath stand for in consciousness?

16. What is the modern "golden calf" that most men worship?

17. Name some of the evil results of the error of money worship.

18. Would the doing away with money entirely solve the problem? What is the solution?

19. What "little but mighty" idea in consciousness is symbolized by David?

20. What are the weapons of this David that slay the giant fear?

www.ingramcontent.com/pod-product-compliance
Lightning Source LLC
Chambersburg PA
CBHW081015040426
42444CB00014B/3220